Introduction to

Industrial Gas Cleaning

Introduction to

Industrial Gas Cleaning

F. A. L. Dullien

Department of Chemical Engineering
University of Waterloo
Waterloo, Ontario, Canada

Academic Press, Inc.

Harcourt Brace Jovanovich, Publishers

San Diego New York Berkeley Boston
London Sydney Tokyo Toronto

ACADEMIC PRESS, INC.
San Diego, California 92101

United Kingdom Edition published by
ACADEMIC PRESS LIMITED
24-28 Oval Road, London NW1 7DX

Library of Congress Cataloging-in-Publication Data

Dullien, F. A. L.
Introduction to industrial gas cleaning.

Includes index.
1. Air—Purification—Equipment and supplies.
2. Flue gases—Purification. 3. Gases—Cleaning.
4. Fume control. I. Title.
TD889.D85 1988 628.5′32 88-3326
ISBN 0-12-223652-1 (alk. paper)

PRINTED IN THE UNITED STATES OF AMERICA
88 89 90 91 9 8 7 6 5 4 3 2 1

To the memory of Werner Strauss

Contents

3 Gravity, Momentum, and Centrifugal Separators

4 Aerodynamic Capture of Particles

5 Gas Filtration

6 Wet Scrubbers

7 Electrostatic Precipitators

8 Gaseous Pollutants

9 Source Testing

Preface

Introduction to Industrial Gas Cleaning evolved from a set of class notes for teaching at the University of Waterloo, which were revised over a period of almost ten years. The course, "Air Pollution Control," is offered to fourth year chemical engineering students as an elective in the second semester. These students have already had physics, fluid mechanics, chemistry, and physical chemistry, which may be regarded as the most useful background subjects for an easy grasp of the physical and chemical principles involved in the functioning and design of gas cleaning equipment.

Throughout the book, great care has been taken to define the terms used, to explain the physical and chemical principles, to state the assumptions made, and to go through the mathematical developments in detail. This approach to the subject has resulted in a rather rigorous treatment, which at the same time is readable, as attested by many streams of chemical engineering students who used the class notes that formed the basis of this book. The text can be used in a second semester junior or senior course in air pollution control or industrial gas cleaning for engineering or science students. It is also useful as a reference in a graduate level course or for engineers, scientists, technicians, managerial personnel, or government officials involved in air pollution control and/or gas cleaning problems.

The teaching of air pollution control at the University of Waterloo was introduced in 1974 at which time there were very few, if any, suitable textbooks. The first edition of the reference text *Industrial Gas Cleaning* by W. Strauss was chosen at that time as the main source for this course, owing to its solid technical content. Some of the material presented in that book was gradually made more suitable for teaching purposes by working out numerical examples, deriving formulas, and in doing so, stating and discussing relevant assumptions and elaborating on the physical and chemical principles involved. Simultaneously, a great deal of material from other sources was incorporated into the class notes and then into this book. Throughout the book, the time-honored principles applied in engineering textbooks have been adhered to. Presentation of purely theoretical approaches that do not have direct impact on practical design has been kept to a minimum.

Thanks are due to my colleague Dr. D. R. Spink, who initiated the air pollution control course and with whom I often exchanged views on the subject matter of this course, and to the hundreds of chemical engineering students who took the course and who contributed to the development of this text by their questions and criticisms. The professional help of Dianne E. Taylor-Harding in correcting and completing the references is gratefully acknowledged. For the illustrations, thanks are due to Mr. Rinze Koopmans, and for the excellent typing of the final manuscript, the credit goes to Mrs. Susie Bell. The invaluable assistance of Ann Collins in compiling the index is gratefully acknowledged.

1

Introduction

1.1 Sources of Air Pollutants

Air pollution may be either natural or man-made. Recently, there has been a tendency by the media to underemphasize the extent of natural air pollution, which, nevertheless, is very great. Forest fires, volcanic eruptions, ocean spray, biological decay, dust storms, pollens, etc., represent vast sources of air pollutants. Natural pollution, on occasion, has caused major environmental readjustments, because it has often proved to be disastrous to plant and animal life; however, life on the planet went on all the same. The effects of man-made air pollution may either be limited to an area or region of varying but finite size or may have far-reaching consequences for life on the entire globe by threatening to alter the global climate. Air pollution problems are mostly of the limited character, but as they arise near densely inhabited areas, they have both harmful and irritating effects.

Most of the air pollution originates from combustion processes, which include power stations, space heating, and motor vehicles. Smelting and other mineral processing operations are important contributors as is the manufacture of steel, coke, lime, ceramics, and cement. Mining and quarrying and chemical industries are also important contributors. Special problems are presented by noxious odors emanating from certain chemical plants as well as from operations such as blood drying, offal rendering, tanning, food processing, glue manufacturing, and fish meal preparation, pig and cattle transportation, and spreading of manure in the fields.

The contributions to air pollution by various sectors as well as the nature and the emission factors of various air pollutants in the United States are

Table 1.1 Estimated Air Pollution Emissions in the United States in 1975[a]

	10^6 (ton/year) and percentage					
	Particulates	Sulfur oxides	Nitrogen oxides	Hydrocarbons	Carbon monoxide	Total
Transportation	1.3 (0.6)	0.8 (0.4)	10.7 (5.3)	11.7 (5.8)	77.4 (38.3)	101.9 (50.4)
Fuel combustion, stationary	6.6 (3.3)	26.3 (13.0)	12.4 (6.1)	1.4 (0.7)	1.2 (0.6)	47.9 (23.7)
Industrial	8.7 (4.3)	5.7 (2.8)	0.7 (0.3)	3.5 (1.7)	9.4 (4.6)	28.0 (13.8)
Solid waste disposal	0.6 (0.3)	<0.1 (~0)	0.2 (0.1)	0.9 (0.4)	3.3 (1.6)	5.0 (2.5)
Miscellaneous	0.8 (0.4)	0.1 (~0)	0.2 (0.1)	13.4 (6.6)	4.9 (2.4)	19.4 (9.6)
Total	18.0 (8.9)	32.9 (16.3)	24.2 (12.0)	30.9 (15.3)	96.2 (47.6)	202.2 (100.0)

[a]EPA (1976).

shown in Table 1.1 (EPA, 1976). This information is not up-to-date, but is nevertheless typical and informative. The total amount of air pollutants generated in North America today can be estimated in excess of 200×10^6 tons a year. This closely corresponds to the mass of a 1000-ft-high mountain occupying an area of one-tenth of a square mile. As the bulk of the pollutants consists of gases, the amount of pollutants is more impressively stated as gas volume: it would cover an area of 14×14 km with a 1-km-high gas blanket.

1.2 Classification of Air Pollutants

There are several different classifications of air pollutants, all of which may be valid. The broad classification given here is used throughout this text. According to this, air pollutants may be subdivided into (1) particulate matter, (2) gases and vapors, and (3) odorous substances.

Particulate matter may consist either of liquid droplets or solid particles. Several different classifications of airborne particulate matter exist, which sometimes result in ambiguous terminology. For example, the following two classifications of terminology are in use:

Industrial terminology		
Coarse dust	> 100 μm in diameter	Solid
Fine dust	1–100 μm in diameter	Solid
Spray	> 10 μm in diameter	Liquid
Mists	1–10 μm in diameter	Liquid
Fumes and smoke	0.001–1 μm in diameter	Solid or liquid
Meteorological terminology		
Rain	> 500 μm in diameter	Liquid
Drizzle	200–500 μm in diameter	Liquid
Mist	80–200 μm in diameter	Liquid
Clouds and fog	2–80 μm in diameter	Liquid
Smog	0.1–2 μm in diameter	Liquid

It is apparent that in industry the term "mist" means a very fine fog-like spray, whereas in the jargon of meteorology, it means a fine rain that is coarser than fog.

All airborne particulates in the range of a particle size of 0.001–100 μm form aerosols. The definition of aerosol is a "disperse system with a gas-phase medium and a solid or liquid disperse phase." The scientific background for the separation of particulates from gases is provided by the science of aerosol mechanics.

The gaseous and vapor-like pollutants may be subdivided into these categories:

- Sulfur-containing compounds (i.e., SO_2 and H_2S) that oxidize into SO_3, H_2SO_4, and MSO_4 (M, metal)
- Nitrogen-containing compounds (i.e., NO and NH_3) that oxidize into NO_2 and MNO_3 (M, metal)
- Organic compounds (i.e., C_1–C_5) that oxidize into ketones, aldehydes, and acids
- Oxides of carbon (i.e., CO)
- Halogens and halides (i.e., Cl_2, HCl, HF, and F_2)

Odorous substances may have a variety of different origins. They may be present in extremely low concentrations where they may defy any attempt to identify their chemical nature.

1.3 Effects of Air Pollutants

Air pollutants have many harmful effects that may be categorized in a number of different ways, for example:

1. Reduction of atmospheric visibility
2. Interference with the weather (in a subtle way)
3. Materials damage (metals, building materials, protective coatings, etc.)
4. Reduction of indoor air quality
5. Damage to vegetation and life in lakes (smelters and power plants, smog, etc.)
6. Biological damage (pathological effects, carcinogenic effects, etc.)

Volumes have been written in the technical literature on the effects of air pollutants. The interested reader is referred to the specialized literature [e.g., Stern (1976)].

In the light of the numerous harmful effects of air pollutants, air pollution control is obviously necessary.

1.4 Air Pollution Legislation and Enforcement

The extent of air pollution control, however, is something to be determined in each particular situation, and this poses a lot of problems. Whereas the moral obligation on the part of industry to control its emissions is undisputed by all responsible management, it is the first responsibility of the management to assure survival and success of the company in a highly competitive economy. Air pollution control, being expensive, adds to the cost of the products, and so it may jeopardize the competitive position of a company. This is one of the reasons why air pollution control cannot be left up to individual companies but must be legislated, and the law enforced evenhandedly. Air quality standards (AQS) have been established to define levels of air quality by certain air quality criteria (AQC) to protect the public health. Secondary AQS define levels of air quality to protect the public welfare from any adverse effects of a pollutant. Different countries, even different states in the United States, have different AQS that are broken down to the level of various pollutants. The reader may find more detailed information on this subject in the current literature (e.g., Licht, 1980; Strauss, 1975). The enforcement of air pollution legislation is generally done by governmental air pollution control agencies, which work with both the industry and the public. Complaints about air pollution problems usually originate with the public, who through political pressure has the power to force the closing of an operation unless it is satisfied with the degree of air pollution control applied there.

1.5 Means of Air Pollution Control

There are essentially three broad categories of methods of air pollution control, classified on the basis of the philosophy of the approach to the problem of prevention of air pollution. These are presented in the order of increasing ultimate effectiveness.

First, historically the idea was both to move the pollutants as far from the plant as possible and dilute them as much as possible. This was accomplished by using tall stacks, some of which are as high as 350 m. While tall stacks usually do the job required of them, they are expensive

Figure 1 Schematic functioning of air pollution control.

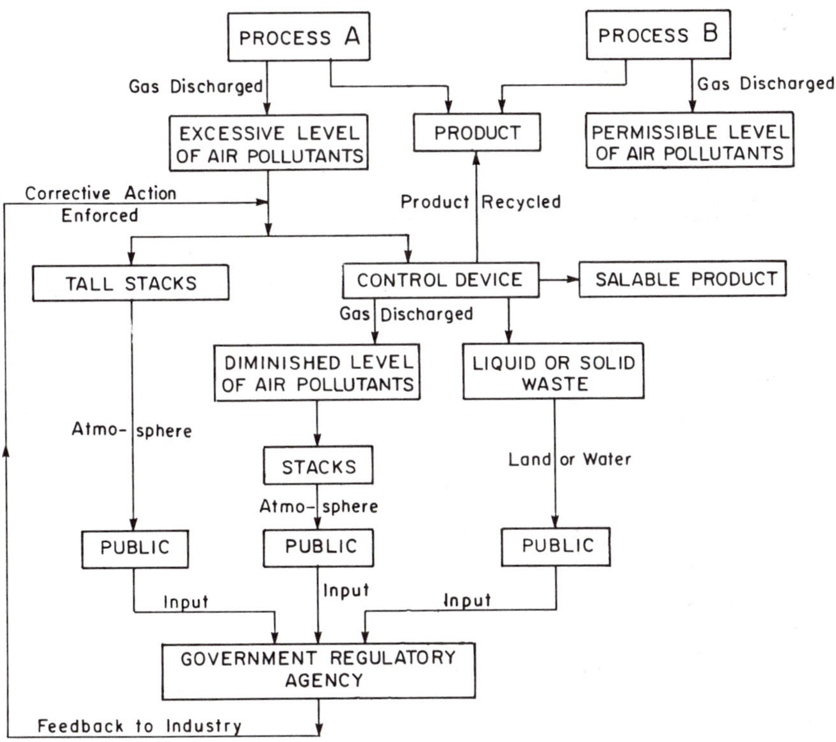

and also simply export the pollution problem somewhere else, where it may or may not cause damage, as the case may be. The tall stacks of power plants emitting SO_2 have been responsible for the occurrence of "acid rain," sometimes thousands of kilometers removed from the stacks.

The second approach consists of the use of the control devices, the purpose of which is to (partially) remove the pollutant, once formed, from the industrial gas before the gas is discharged to the atmosphere. The bulk of this textbook deals with the physical–chemical principles involved in the operation and design methods of control devices. The pollutant, once removed, if not noxious or otherwise damaging to the environment, can be discharged into a body of water, dumped or used as landfill. The pollutant sometimes represents a portion of the product, and after collection it is recycled and added to the product. There is always an incentive to turn the pollutant into a salable product in order to be able to pay for the cost of the air pollution control system.

Third, alternate processes to obtain the desired product without causing any air pollution should be listed as potentially the best way of handling the problem. The alternate processes, however, must be competitive or else the cost aspect cancels any possible advantage these approaches may promise to offer. The scheme of air pollution control is depicted diagrammatically in Fig. 1.1.

1.6 Selection of the Control Device

A survey of the table of contents of this text indicates the existence and availability of a large number of different air pollution control equipment, many of which work on the basis of different principles. A large number of factors must be considered before selecting the type of equipment that is most suitable for the particular air pollution problem to be solved. In addition to the first two obvious criteria (i.e., pollutant removal and cost effectiveness), there are many less obvious but equally important aspects of the process/device to be scrutinized. A very important decision to make is whether to use a "wet" or "dry" process. Another important aspect is whether both particulates and gaseous pollutants need to be removed simultaneously. More often than not a number of different control devices need to be used, one after the other, forming a system, in order to achieve the desired cleaning effect.

The starting point in tackling the air pollution control problem is a good knowledge of the nature and the amount of the pollutants present, obtained by source sampling and subsequent analysis of the sample. The temperature, the humidity, and the dewpoint, as well as the quantity of the gas to be treated, must be known. Start-ups, shutdowns, and possible fluctuations in the gas flow and in the pollutant loading must be taken into consideration. Materials of construction must be selected on the basis of temperature, corrosion, and erosion hazards. In addition to the cost of purchase and installation of the system, operating expenses must also be considered. A very important contributor to the operating expenses may be the pressure drop in the system. Another contributor is the cost of maintenance. Trouble-free operation also translates into savings. Safety aspects represent another very important point to consider. Often it is necessary to build a pilot plant before manufacturing the full-scale unit.

Air pollution, in general, and air pollution engineering, in particular, have been covered extensively in the literature. In addition to the textbooks already mentioned in this chapter, the reader may find useful, although inevitably incomplete, this list of reference texts: Hesketh, 1979; Theodore and Bounicore, 1976; Satriana, 1981; Noll and Duncan, 1973; Calvert and Eglund, 1984; Henzel *et al.*, 1982; Storch *et al.*, 1979; Cooper and Alley, 1986; Schifftner and Hesketh, 1983; Cheremisinoff and Young, 1975, 1977; Jimeson and Spindt, 1973; Young and Cross, 1982; Ogawa, 1984; Stern *et al.*, 1984; Friedlander, 1977; Fuchs, 1964; Hidy and Brock, 1970; Sanders, 1979; Liu, 1976.

References

Calvert, S., and Eglund, H. M., eds. (1984). "Handbook of Air Pollution Technology." Wiley, New York.

Cheremisinoff, P. N., and Young, R. A. (1975). "Pollution Engineering Practice Handbook." Ann Arbor Science, Ann Arbor, Michigan.

Cheremisinoff, P. N., and Young, R. A. (1977). "Air Pollution Control and Design Handbook," Parts 1 and 2. Dekker, New York.

Cooper, C. D., and Alley, F. C. (1986). "Air Pollution Control: A Design Approach." PWS Publ., Boston, Massachusetts.

EPA (1976). "National Air Quality and Emissions Trends Report, 1975." Preliminary data, EPA-450/1-76-002.

Friedlander, S. K. (1977). "Smoke, Dust and Haze—Fundamentals of Aerosol Behavior." Wiley, New York.

Fuchs, N. A. (1964). "The Mechanics of Aerosols" (translated from the Russian). Macmillan, New York.

Henzel, D. S., Laseke, B. A, Smith, E. O., and Swenson, D. O. (1982). "Handbook for Flue Gas Desulfurization Scrubbing with Limestone." Noyes Data Corp., New Jersey.

Hesketh, H. E. (1979). "Air Pollution Control." Ann Arbor Science, Ann Arbor, Michigan.

Hidy, G. M., and Brock, J. R. (1970). "The Dynamics of Aerocolloidal Systems." Pergamon, Oxford.

Jimeson, R. M., and Spindt, R. S., eds. (1973). "Pollution Control and Energy Needs," Advances in Chemistry Series 127. Amer. Chem. Soc., Washington, D.C.

Licht, W. (1980). "Air Pollution Control Engineering." Dekker, New York.
Liu, B. Y. H., ed. (1976). "Fine Particles—Aerosol Generation, Measurement, Sampling, and Analysis." Academic Press, New York.
Noll, K., and Duncan, T. (1973). "Industrial Air Pollution Control." Ann Arbor Science, Ann Arbor, Michigan.
Ogawa, A. (1984). "Separation of Particles from Air and Gases," Parts 1 and 2. CRC Press, Boca Raton, Florida.
Sanders, P. A. (1979). "Handbook of Aerosol Technology," 2nd Ed. Van Nostrand-Reinhold, Princeton, New Jersey.
Satriana, M., ed. (1981). "New Developments in Flue Gas Desulfurization Technology." Noyes Data Corp., New Jersey.
Schifftner, K. C., and Hesketh, H. E. (1983). "Wet Scrubbers." Ann Arbor Science, Ann Arbor, Michigan.
Stern, A. C., ed. (1976). "Air Pollution," Vol. 1–5. Academic Press, New York.
Stern, A. C., Boubel, R. W., Turner, O. B., and Fox, D. L. (1984). "Fundamentals of Air Pollution," 2nd Ed. Academic Press, New York.
Storch, O., *et al.* (1979). "Industrial Separators for Gas Cleaning." Elsevier, Amsterdam.
Strauss, W. (1975). "Industrial Gas Cleaning," 2nd Ed. Pergamon, Oxford.
Theodore, L., and Bounicore, A. J. (1976). "Industrial Air Pollution Control Equipment for Particulates." CRC Press, Cleveland, Ohio.
Young, R. A., and Cross, Jr., F. L., eds. (1982). "Specifying Air Pollution Control Equipment." Dekker, New York.

2

Fluid Resistance to Particle Motion

2.1 Introduction

The problem of particle removal from stack gases always involves a displacement of the particles relative to the gas stream, that is, there is a relative velocity between the particles and the gas. Under these conditions, the gas exerts a friction or drag force on the particles. For example, if a stationary particle is placed in a gas stream, the particle will be accelerated by this force. On the other hand, if a particle is catapulted into a still gas, the fluid resistance will tend to slow down the particle. In both cases, the same type of friction force is active between the particle and the gas, hence it is customary to refer to fluid resistance to particle motion in both instances. It is essential to be familiar with the laws that determine the magnitude of the fluid resistance to the motion of particles under a variety of different conditions in order to be able to calculate the trajectory of the particles. In all of the different types of equipment in use for the removal of particulates from gas streams, the particles are captured by solid or liquid collector surfaces, hence a knowledge of the particle trajectory is essential for the understanding and the design of this kind of equipment. This is the reason for starting our study of air pollution control equipment for particulates with the fluid resistance to particle motion.

2.2 Spherical Particles in Steady Motion

Although real particles are very seldom exactly spherical, most of the detailed experimental study and the theoretical calculations were performed for spherical particles, because a sphere is the simplest shape a particle can have. The behavior of particles other than spheres are usually referred to the behavior of a so-called *equivalent sphere*.

The equivalent sphere diameter is the diameter of a spherical object that behaves in some special respect—which must be clearly defined—exactly as the nonspherical particle does. For example, the equivalent sphere can have the same surface area, the same volume, or the same surface-to-volume ratio as the nonspherical particle in question. Another equivalent sphere diameter is the diameter of the sphere that has the same friction drag as the nonspherical particle in question. We are going to discuss these in greater detail later in this chapter.

It is customary to analyze the motion of particles in stagnant fluid and to assume that the resistance to motion will remain the same under flow conditions. The theory of particle motion relative to flowing fluids is not developed sufficiently to permit accurate predictions, although much research in this area is currently under way. It has been recognized that the fluid resistance to particle motion is, in principle, different, depending on the rate of acceleration of the particle with respect to the fluid. First, we discuss the simplest case of steady motion, that is, zero rate of acceleration of the particle.

Stokes' Law

The resistance to particle motion exerted by the fluid depends on the so-called particle Reynolds number (Re):

$$\mathrm{Re} = ud\rho/\mu \tag{2.1}$$

where u is the relative velocity of the particle with respect to the fluid (or the relative velocity of the fluid with respect to the particle), d is the diameter of the particle, ρ is the density, and μ is the viscosity of the fluid.

For very small values of the Reynolds number, that is, values less than 0.1 (so-called creeping flow), the fluid resistance (or friction force), F, to the steady motion of a sphere, is given by Stokes' law:

$$F = 3\pi\mu du \tag{2.2}$$

This equation expresses the fact that the friction force F is directly proportional to the viscosity of the fluid, the diameter of the particle, and the relative velocity between the particle and the fluid. Its detailed derivation and discussion can be found in (Lamb, 1945; Happel and Brenner, 1965; Bird *et al.*, 1960).

In creeping flow (often referred to as "viscous" flow), inertial effects are completely neglected relative to viscous forces. As a result, the streamline

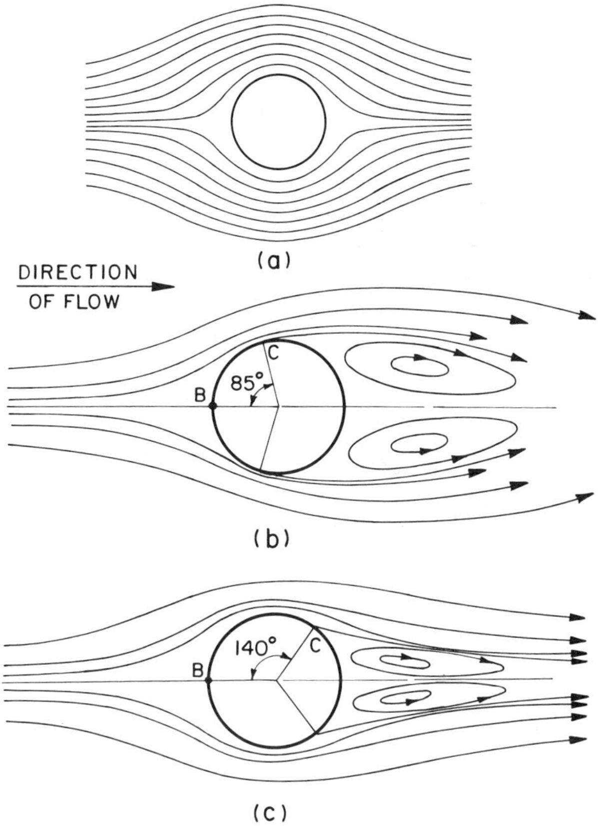

Figure 2.1 (a) Flow past single sphere in creeping flow, and showing separation and wake formation; (b) laminar flow in boundary layer; (c) turbulent flow in boundary layer. B, Stagnation point; C, separation point.

pattern is symmetrical with respect to the direction of flow, as shown in Fig. 2.1(a).

Drag Coefficient

In engineering practice, the drag coefficient C_D is most commonly used and is defined as

$$C_D = F/A(\rho u^2/2) \tag{2.3}$$

where A is the area of the particle projected on a plane perpendicular to the

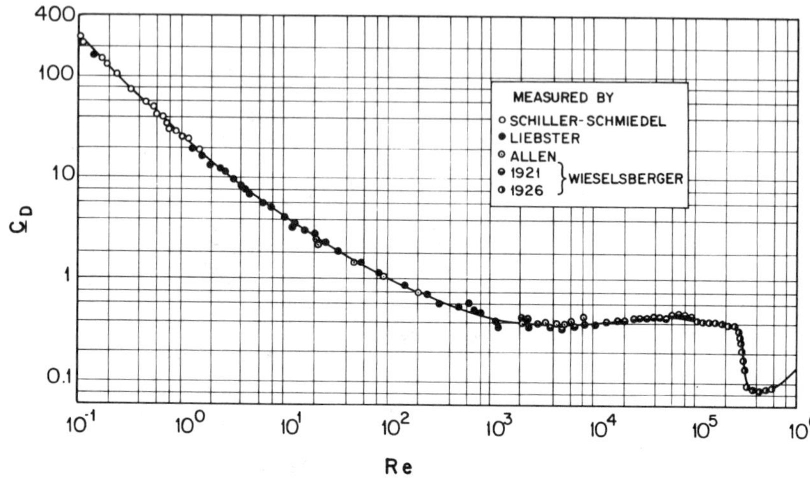

Figure 2.2 Drag coefficient for spheres as a function of the Reynolds number.

direction of motion of the particle. For example, for spheres we have

$$A = \pi d^2/4 \tag{2.4}$$

Combining Eqs. (2.1)–(2.4), we obtain

$$C_D = 24/\text{Re} \tag{2.5}$$

Equation (2.5) shows that for the range of validity of Stokes' law (for engineering purposes it is often used up to Re $\approx$ 2), the drag coefficient is inversely proportional to the Reynolds number as can be seen in Fig. 2.2, where the relationship C_D versus Re is plotted on a log–log diagram. Part of this plot is approximately a straight line of slope -1. For higher values of the Reynolds number, the slope of the curve decreases (so-called transition region). In the range of 10^3–10^5 Re, called Newton's law region, the value of the drag coefficient is practically constant and lies between 0.38 and 0.5.

Beyond the range of Stokes' law, a boundary layer forms, originating at point B. The fluid separates, and a wake is also formed, as shown in Fig. 2.1(b); at even higher Reynolds numbers, turbulence occurs in the boundary layer, C_D drops from 0.45 to 0.10 (Fig. 2.2), and the separation points move toward the rear of the sphere, as shown in Fig. 2.1(c).

It is useful to try to express the relationship between the drag coefficient and the Reynolds number in analytical form. Some of the more important equations obtained by various authors are listed here:

$$C_D = (24/\mathrm{Re})(1 + 3\,\mathrm{Re}/16), \qquad (\text{up to Re} \simeq 1)$$
Oseen (see Lamb, 1945, p. 612) (2.6)

$$C_D = 10/\mathrm{Re}^{1/2}, \qquad (30 \le \mathrm{Re} \le 300) \qquad \text{Allen (1900)} \tag{2.7}$$

$$C_D = 24/\mathrm{Re} + 4/\mathrm{Re}^{1/3}, \qquad (3 \le \mathrm{Re} \le 400)$$
Klyachko (see Fuchs, 1964, p. 33) (2.8)

$$C_D = (24/\mathrm{Re})(1 + 0.15\,\mathrm{Re}^{0.687}), \qquad (0.5 \le \mathrm{Re} \le 800)$$
Schiller and Naumann (1933) (2.9)

$$C_D = (24/\mathrm{Re})(1 + 0.197\,\mathrm{Re}^{0.63} + 0.0026\,\mathrm{Re}^{1.38}), \qquad (1 \le \mathrm{Re} \le 100)$$
Langmuir and Blodgett (1946) (2.10)

$$C_D = b_1/\mathrm{Re}^n, \qquad \text{McCabe and Smith (1976)} \tag{2.11}$$

where for $2 \le \mathrm{Re} \le 500$ (transition range), $b_1 = 18.5$ and $n = 0.6$, and for $500 \le \mathrm{Re} \le 200{,}000$ (Newton's law range), $b_1 = 0.44$ and $n = 0$.

Terminal Settling Velocity

The drag coefficient versus the Reynolds number plot or, alternatively, Eqs. (2.6)–(2.11) are used to solve the following typical problem: Given a particle of a certain size, shape, and density in a fluid of a given density and viscosity, where the particle is subjected to a known driving force, what is the steady state velocity, often referred to as the terminal velocity, of the particle? The terminal velocity of the particle is denoted by u_t.

The solution to this problem is obtained by setting up a steady state force balance, that is, in the steady state, the driving force G is equal to the resisting, or drag force, F. It is noted that when the forces cancel there is no acceleration, and no acceleration corresponds to steady motion of the particle. Hence, we write

$$F = G \tag{2.12}$$

Combining Eq. (2.3) with Eq. (2.12), the general expression for the terminal velocity of the particle is obtained:

$$u_t = \sqrt{\frac{2G}{A\rho C_D}} \tag{2.13}$$

We note that Eq. (2.13) is subject to the restriction of steady motion, but it applies equally to particles of any arbitrary shape, any value of the Reynolds number, and any type of driving force G.

If the particle is spherical and the Reynolds number is low enough, so that Stokes' law is valid, we can combine Eqs. (2.4) and (2.5) with Eq. (2.13) to obtain this expression for the terminal velocity:

$$u_t = G/3\pi d\mu \tag{2.14}$$

Now, in the special case when the driving force is gravity, G can be expressed as:

$$G = (d^3\pi/6)(\rho_p - \rho)g \tag{2.15}$$

where ρ_p is the particle density, and g is the gravitational acceleration constant. The fluid density ρ was subtracted from the particle density ρ_p to allow for the buoyancy effect of the fluid. In most situations where the fluid is air, this correction is negligible.

Combining Eqs. (2.14) and (2.15), an important expression for the terminal settling velocity of a spherical particle in the region of validity of Stokes' law is obtained[1]:

$$u_t = \frac{d^2(\rho_p - \rho)g}{18\mu} \tag{2.16}$$

Often, Stokes' law is not valid because the value of the Reynolds number exceeds 0.1 considerably. In such cases, we may combine Eq. (2.13) with Eqs. (2.4) and (2.15) to obtain for a spherical particle:

$$u_t = \sqrt{\frac{4d(\rho_p - \rho)g}{3\rho C_D}} \tag{2.17}$$

It is noted that C_D is a function of the Reynolds number and, hence, the terminal settling velocity u_t. Therefore, Eq. (2.17) is an implicit equation for u_t, which would normally require trial-and-error solution. In order to avoid this, the artifice devised by von Kármán is used: for spherical particles, Eqs. (2.13) and (2.4) yield

$$C_D = \frac{8G}{\pi\rho u_t^2 d^2} \tag{2.18}$$

Multiplying C_D by Re^2, u_t is eliminated from the resulting expression:

$$C_D\,\mathrm{Re}^2 = \frac{8G}{\pi\rho u_t^2 d^2}\,\frac{u_t^2 d^2\rho^2}{\mu^2} = \frac{8G\rho}{\pi\mu^2} \tag{2.19}$$

[1]The general form of Eq. (2.16) includes the Cunningham correction factor [see Eq. (2.46)], i.e., $u_t = [Cd^2(\rho_p - \rho)g]/18\mu$.

If the driving force is gravity, use of Eq. (2.15) results in

$$C_D \, \mathrm{Re}^2 = 4(\rho_p - \rho) d^3 g / 3\mu^2 \tag{2.20}$$

Note that all the quantities on the right-hand side of Eq. (2.20) are known parameters; hence one does not have to know the values of C_D and Re in order to evaluate Eq. (2.20). However, one can replot the graph of Fig. 2.2 in the form Re versus $C_D \, \mathrm{Re}^2$, whence Re may be obtained by evaluating the right-hand side of Eq. (2.20). From the value of the Reynolds number obtained in this way, one calculates the value of the terminal velocity directly by Eq. (2.1).

Instead of giving the relationship Re versus $C_D \, \mathrm{Re}^2$ in graphical form, one can present the same in analytical form. This has been done by Davies (1945):

$$\mathrm{Re} = \frac{C_D \, \mathrm{Re}^2}{24} - 2.3363 \times 10^{-4} (C_D \, \mathrm{Re}^2)^2 + 2.0154 \times 10^{-6} (C_D \, \mathrm{Re}^2)^3$$
$$- 6.9105 \times 10^{-9} (C_D \, \mathrm{Re}^2)^4, \qquad (\mathrm{Re} \le 4;\ C_D \, \mathrm{Re}^2 \le 134) \tag{2.21}$$

and

$$\log_{10} \mathrm{Re} = -1.29536 + 0.986 \log_{10}(C_D \, \mathrm{Re}^2)$$
$$- 0.046677 \left[\log_{10}(C_D \, \mathrm{Re}^2)\right]^2$$
$$+ 0.0011235 \left[\log_{10}(C_D \, \mathrm{Re}^2)\right]^3$$
$$(3 \le \mathrm{Re} \le 10{,}000;\ 100 \le C_D \, \mathrm{Re}^2 \le 4.5 \times 10^7) \tag{2.22}$$

Table 2.1 contains useful information on the maximum diameters of spheres of various densities for which the terminal falling velocities can be calculated by Stokes law or Eq. (2.21).

An alternate method of evaluating Eq. (2.17) is offered by combining it with Eq. (2.11), giving (McCabe and Smith, 1976)

$$u_t = \left[\frac{4 g d^{(1+n)} (\rho_p - \rho)}{3 b_1 \mu^n \rho^{(1-n)}} \right]^{1/(2-n)} \tag{2.23}$$

The appropriate values of b_1 and n can be decided on by evaluating the parameter K, defined as (McCabe and Smith, 1976)

$$K = d \left[\frac{g \rho (\rho_p - p)}{\mu^2} \right]^{1/3} \tag{2.24}$$

For the transition range, $3.3 \le K \le 43.6$; and for Newton's law range, $43.6 \le K \le 2{,}360$.

Table 2.1 The Maximum Sizes of Spheres for Which Terminal Falling Velocities Can Be Calculated by Stokes' Law[a, b]

Density of sphere ($kg\ m^{-3}$)	Maximum diameter for Stokes' law Within 10% (μm)	Within 5% (μm)	Within 1% (μm)	Maximum diameter with Eq. 2.21 (μm)
200	132	100	57	240
400	105	79	45	191
800	83	63	36	152
1,000	77	59	34	141
2,000	61	46	27	112
4,000	48	37	21	89
6,000	42	32	18	78
8,000	38	29	17	70
10,000	36	27	15	65
12,000	34	25	15	62
Re	0.82	0.38	0.074	4
$C_D\ Re^2$	21.9	9.60	1.80	133.6

[a] The spheres are falling through air at 20°C and 100 kPa pressure ($\mu = 1.82 \times 10^{-5}$ Pa s; $\rho = 1.205\ kg\ m^{-3}$). See Eqs. (2.16) and (2.21).
[b] Reprinted with permission from Strauss (1975).

2.3 Fluid Resistance to Accelerating Spheres

Whenever the velocity of the particle is not constant, the particle is either accelerating or decelerating (i.e., accelerating with a negative sign). The drag force on the accelerating particle is larger than that on a particle moving with the same velocity in steady state. The higher the rate of acceleration of the particle with respect to the fuid, the greater is the drag coefficient for the same value of the Reynolds number. The variation of the drag coefficient with the Reynolds number for various rates of acceleration of the particle is shown in Fig. 2.3 (Torobin and Gauvin, 1959, 1960, 1961). Note that this diagram is drawn on arithmetic coordinates and it covers only a limited range of the Reynolds number, from a few thousand to about 50,000.

A good example is the acceleration of a particle dropped in still air. At time zero the particle is accelerating at the full gravitational acceleration g. As the velocity of the particle increases and the friction force builds up, the particle accelerates less and less, and finally, when the terminal velocity u_t has been reached, the acceleration is equal to zero.

A force balance serves as the basis for the calculations also under unsteady state conditions. In this case the force balance expresses the fact

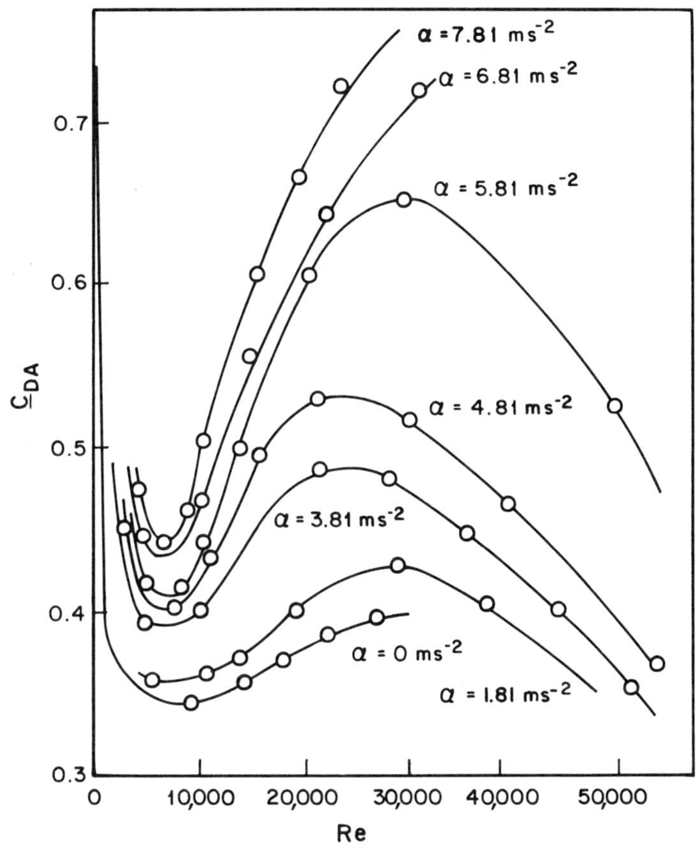

Figure 2.3 Modified drag coefficient C_{DA} for accelerating spheres falling in air. [Redrawn by Torobin and Gauvin (1959, 1960, 1961) based on experiments in Lunnon (1928).]

that the driving force $\mathbf{G}$ is greater than the resisting force $\mathbf{F}_A$, and the difference between the two forces is the net force acting on the particle, resulting in acceleration, that is, one writes

$$\mathbf{G} - \mathbf{F}_A = m\mathbf{a} \tag{2.25}$$

where m is the mass and $\mathbf{a}$ is the acceleration of the particle.

In the case of nonsteady motion of particles, special attention has been given to the range of low Re where, in the case of steady motion, Stokes' law is valid. In the case of unsteady motion of a sphere in a stagnant fluid, the friction force is given by a modification of Stokes' law (Fuchs, 1964):

$$F_A = 3\pi\mu du + \left(\frac{\pi}{12}\right)\rho d^3 a + \left(\frac{3}{2}\right) d^2\sqrt{\pi\mu\rho}\int_0^t \frac{du}{dt'}\frac{dt'}{\sqrt{t-t'}} \tag{2.26}$$

The first term in Eq. (2.26) is the resistance to a sphere in steady motion (Stokes drag), the second term represents the resistance of an ideal fluid to the accelerating motion of a sphere ($\mu = 0$), and the third term expresses that part of the resistance which is due to setting the fluid into motion. It has been shown that the third term in Eq. (2.6) may be neglected without introducing any appreciable error under all conditions, regardless of the size of the particle. The relative importance of the second term in Eq. (2.26) is obviously a function of fluid density and particle size. The smaller the particle, the more negligible the second term becomes, compared with the first term. In all the calculations made in this text, both the second and third terms are always neglected.

For the special case of a small particle introduced into a stationary fluid, and accelerating in the gravity field, the combination of Eqs. (2.25), (2.26), and (2.15) results in the differential equation

$$\frac{du}{dt} + \frac{u}{\tau} - g = 0 \tag{2.27}$$

where the second and third terms of Eq. (2.26) have been neglected. In Eq. (2.27), τ is the relaxation time of the particle, that is,

$$\tau = m/3\pi\mu d \tag{2.28}$$

which becomes, for the special case of a spherical particle,[2]

$$\tau = d^2\rho_p/18\mu \tag{2.29}$$

Integrating Eq. (2.27), using as the initial condition the particle at rest, one obtains this expression for the particle velocity as a function of time:

$$u = \tau g[1 - \exp(-t/\tau)] \tag{2.30}$$

Integrating Eq. (2.30) once more, the next expression is obtained for the distance x covered by the falling particle as a function of time.

$$x = \tau g t + \tau^2 g[\exp(-t/\tau) - 1] \tag{2.31}$$

It should be noted that these equations are valid only at very low values of the Reynolds number, that is, when in steady motion Stokes' law is valid. Indeed it is evident that for large values of t Eq. (2.30) becomes

$$u_t = \tau g \qquad \text{(large time)} \tag{2.32}$$

which is identical to Eq. (2.16).

[2] General form of Eq. (2.27) includes the Cunningham correction factor [see Eq. (2.46)], i.e., $\tau = Cd^2\rho_p/18\mu$.

Another example for a particle in unsteady motion is when an initially stationary particle is picked up by a flowing fluid. In this case, the particle is decelerating with respect to the fluid, because the velocity difference between the particle and the fluid is greatest at the outset, and in the final, steady state it is equal to zero. One must be very careful in the analysis of this kind of problem, because the particle does accelerate with respect to the walls of the duct in which the gas is flowing. One also notes that **G** in Eq. (2.25) is equal to zero in this case.

Whereas for low values of the particle Reynolds number, the same drag coefficient, C_D, may be used also for accelerating particles, for large values of the Reynolds number, drag coefficients determined under accelerating conditions, such as those shown in Fig. 2.3, must be used. Under unsteady state conditions, the drag coefficient C_{DA} is defined analogously as under steady conditions (Torobin and Gauvin, 1959, 1960, 1961):

$$C_{DA} = \frac{F_A}{(1/2)\rho u^2 A} \tag{2.33}$$

It is noted that F_A and C_{DA} depend on a. Whereas at very low Re, the friction force is practically the same both under accelerating conditions and in the steady state, at high Re, F_A becomes significantly greater than F.

Assuming a constant drag coefficient C_{DA}, relationships can be derived (Torobin and Gauvin, 1959, 1960, 1961) for a particle injected into a stationary gas and accelerating under the influence of gravity, between particle velocity and time, and between the distance covered by a particle and time. The assumption of a constant drag coefficient appears, at first sight, to be very poor; nevertheless, it can be shown that the results obtained with the help of these equations are fairly good and relatively insensitive to the value of C_{DA} used.

In order to obtain the relationship between u and t, one proceeds as follows. Equation (2.25) is combined with Eq. (2.33) to give

$$C_{DA}(1/2)\rho u^2 A = G - m(du/dt) \tag{2.34}$$

For the case of a spherical particle, we can use Eq. (2.4), and

$$m = \rho_p(d^3\pi/6) \tag{2.35}$$

Hence, after separating the variables, Eq. (2.34) becomes

$$\int_0^t dt = \frac{4\rho_p d}{3C_{DA}\rho}\int_0^u \frac{du}{8G/\pi d^2 C_{DA}\rho - u^2} \tag{2.36}$$

The integral at the right side of Eq. (2.36) is evaluated as

$$\int \frac{du}{a^2 - u^2} = \int \frac{du}{(a+u)(a-u)} = \frac{1}{2a} \int \frac{(a+u) + (a-u)}{(a+u)(a-u)}\, du$$

$$= \frac{1}{2a} \frac{-(a+u)d(a-u) + (a-u)d(a+u)}{[(a+u)/(a-u)](a-u)^2}$$

$$= \frac{1}{2a} \int \frac{d(a+u)/(a-u)}{(a+u)/(a-u)} = \frac{1}{2a} \int d \ln\left(\frac{a+u}{a-u}\right)$$

where

$$a = (8G/C_{\mathrm{DA}} d^2 \pi \rho)^{1/2}$$

The final result is

$$t = \frac{1}{3} \frac{\rho_{\mathrm{p}} d^2}{\sqrt{C_{\mathrm{DA}} \rho}} \sqrt{\frac{\pi}{2G}} \ln \frac{(2/d)\sqrt{(2G/\pi C_{\mathrm{DA}} \rho)} + u}{(2/d)\sqrt{(2G/\pi C_{\mathrm{DA}} \rho)} - u} \tag{2.37}$$

In order to find the distance covered by the accelerating particle in time t, one has to integrate Eq. (2.37) once more.

$$t = A \ln \frac{B + u}{B - u}$$

Hence,

$$u = B \frac{e^{t/A} - 1}{e^{t/A} + 1} = B \frac{(e^{t/A} - 1)(e^{-t/A} + 1)}{(e^{t/A} + 1)(e^{-t/A} + 1)}$$

$$= B \frac{e^{t/A} - e^{-t/A}}{(e^{t/A} + 1)(e^{-t/A} + 1)} = B\left(\frac{1}{e^{-t/A} + 1} - \frac{1}{e^{t/A} + 1}\right) = \frac{dx}{dt}$$

where

$$B = \frac{2}{d} \sqrt{\frac{2G}{\pi C_{\mathrm{DA}} \rho}}$$

and

$$A = \frac{1}{3} \frac{\rho_{\mathrm{p}} d^2}{\sqrt{C_{\mathrm{DA}} \rho}} \sqrt{\frac{\pi}{2G}}$$

Rearrangement and subsequent integration of Eq. (2.37) gives

$$x = B\left(\int_0^t \frac{dt}{e^{-t/A}+1} - \int_0^t \frac{dt}{e^{t/A}+1}\right) \tag{2.38}$$

$$\frac{dt}{e^{-t/A}+1} = \frac{dt + e^{-t/A}dt - e^{-t/A}dt}{e^{-t/A}+1} = \frac{(1+e^{-t/A})\,dt - e^{-t/A}dt}{e^{-t/A}+1}$$

$$= dt - \frac{e^{-t/A}dt}{e^{-t/A}+1} = dt + A\frac{d(e^{-t/A}+1)}{e^{-t/A}+1}$$

and

$$\int_0^t \frac{dt}{e^{-t/A}+1} = t + A\ln\frac{e^{-t/A}+1}{2}$$

Analogously,

$$\int_0^t \frac{dt}{e^{t/A}+1} = t - A\ln\frac{e^{t/A}+1}{2}$$

Upon evaluating the integrals on the right-hand side of Eq. (2.38), we obtain

$$x = AB\ln\frac{2+e^{-t/A}+e^{t/A}}{4} = AB\ln\left(\frac{e^{-t/2A}+e^{t/2A}}{2}\right)^2 \tag{2.39}$$

Finally,

$$x = 2AB\ln\cosh\frac{t}{2A} = \frac{4}{3}\frac{\rho_p d}{\rho C_{DA}}\ln\cosh\left[\frac{3\sqrt{2C_{DA}\rho G}}{2d^2\rho_p\pi^{1/2}}t\right] \tag{2.40}$$

2.4 Deviations from Stokes' Law When the Particle Radius Is Either about the Same As or Less Than the Mean Free Path

Stokes' law and the empirical relationship between the drag coefficient and the Reynolds number shown in Fig. 2.2 are valid only if $d/2 \gg \lambda$, the mean free path of the gas molecules (Fuchs, 1964):

$$\lambda = \frac{1}{\sqrt{2}\,\pi N\sigma^2} = \frac{\mu}{0.499\rho\bar{u}} \tag{2.41}$$

where N is the number of molecules per cubic centimeter, σ is the molecular diameter, and $\bar{u}$ is the mean speed of the molecules, given by the equation

$$\bar{u} = (8RT/\pi M)^{1/2} \tag{2.42}$$

The order of magnitude of λ in ambient air is 0.01–0.1 μm.

When the particle diameter is very much greater than the mean free path, the particles "see" the gas as a continuum, that is, they "are not aware" of the existence of individual molecules. Classical hydrodynamics treats the fluids as continua, and Stokes' law was derived on this basis. Whenever the particle size is either of the same order of magnitude as or smaller than the mean free path, the molecular nature of the gas makes itself felt, resulting in a different type of interaction between the particle and the gas than the relationship expressed by Stokes' law. Next, we discuss three different cases of deviation from Stokes' law.

CASE 1. $d/2 \ll \lambda$: This is the case of very finely dispersed aerosols, the motion of which is of molecular character. Each aerosol particle behaves as a large molecule as it keeps colliding with the gas molecules. The number of molecules impinging on the surface of a particle at any moment from various directions is not constant, but undergoes fluctuations, resulting in the so-called "Brownian motion," of diffusion of the particles.

When a particle is moving in a certain preferred direction because of a driving force, for example, gravity, then the resistance offered by the gas to the motion of the particle is caused by a greater number of molecules impinging on the frontal surface of the moving particle than from behind. This resistance is proportional to the surface area of the particle or the square of the particle diameter. For $d < 10^{-7}$ cm the resistance is expressed by this formula (Fuchs, 1964):

$$F = [3\pi\mu du/(A + Q)](d/2\lambda) \tag{2.43}$$

A comparison of Eq. (2.43) with Stokes' law [Eq. (2.2)] shows that when the particle radius is much smaller than the mean free path, the resistance offered by the fluid is much less than when the particle radius is greater than the mean free path, that is, when Stokes' law is valid.

CASE 2. $d/2 \gtrsim \lambda$ ($d/2$ approaches λ from above): It can be shown theoretically that as the particle size is decreased from $d/2 \gg \lambda$, and it approaches the mean free path, a velocity discontinuity develops in the gas near the particle surface. As a result, the relative velocity between the particle and the first layer of gas molecules next to the solid surface is not equal to zero, as is the case in ordinary aerodynamics. This effect also exists, in principle, when $d/2 \gg \lambda$, but it becomes noticeable only as $d/2 \to \lambda$. The phenomenon is referred to as "slip," because it appears as if the particle and the adjacent layer of fluid are slipping with respect to each other.

Under these conditions Stokes' law has been modified as (Fuchs, 1964)

$$F = 3\pi\mu du/C \tag{2.44}$$

where C, the Cunningham correction factor (Cunningham, 1910), is given by

$$C = 1 + A(2\lambda/d) \tag{2.45}$$

It is evident from these equations that in the presence of a "slip" the friction force is diminished.

CASE 3. $d/2 \simeq \lambda$ ($d/2$ is of the same order of magnitude as λ): This case has not been treated successfully by theoretical means, but an empirical interpolation formula has been developed that can be written in the same form as Eq. (2.44), where C, the generalized Cunningham correction factor, is now given by (Fuchs, 1964)

$$C = 1 + (2\lambda/d)[A + Q\exp(-bd/2\lambda)] \tag{2.46}$$

Owing to the empirical nature of Eq. (2.46) and also to experimental errors in the data, several different sets of values have been proposed in the literature for the constant parameters, A, Q, and b (e.g., $A = 1.246$, $Q = 0.42$, and $b = 0.87$). Equation (2.44) used with Eq. (2.46) is the most general form of Stokes' law which simplifies, under the appropriate conditions, to the various modified forms of Stokes' law. It can be easily seen from Eq. (2.46) that as $2\lambda/d \gg 1$, $C \to (2\lambda/d)(A + Q)$ (Case 1), as

Table 2.2 Values of Cunningham Correction Factor for Particles from 0.01 to 10 μm and Temperature from 0 to 1500°C[a]

Temperature (°C)	Air Viscosity ($\mu \times 10$ Pa s)	Particle diameter (μm)			
		0.01	0.10	1.0	10.0
0	17.04	20.15	2.64	1.149	1.015
200	25.85	39.84	4.58	1.299	1.0297
400	32.86	59.89	6.55	1.457	1.0450
600	38.80	80.43	8.59	1.626	1.0606
800	44.05	100.99	10.64	1.801	1.0761
1000	48.76	121.5	12.69	1.981	1.0918
1200	53.10	142.4	14.77	2.171	1.1076
1400	57.13	163.3	16.85	2.363	1.1234
1600	60.90	184.2	18.94	2.557	1.1393

[a]Reprinted with permission from Strauss (1975).

$2\lambda/d < 1$, $C \to 1 + (2\lambda/d)A$ (Case 2), and as $2\lambda/d \ll 1$, $C \to 1$ (Stokes' law case). Values of the Cunningham correction factor are tabulated in Table 2.2.

2.5 Fluid Resistance to Particle Motion in the Case of Clouds of Particles

The resistance exerted to the motion of clouds of particles, or to an individual particle contained in a cloud of particles, is generally greater than the resistance exerted to an isolated individual particle under otherwise identical conditions. This effect is important from a practical point of view as in technological operations one seldom deals with isolated particles. It is convenient to distinguish two categories of clouds of particles.

Unbounded Clouds

The cloud of particles is sometimes unbounded, that is, it is not enclosed by walls of a duct or a vessel. Unbounded clouds have a tendency to move as one entity, which includes the air contained between the particles. Therefore, unbounded clouds have been treated as one big bubble of a density only slightly different from that of the surrounding air. For very small values of the Reynolds number corresponding to extremely slow motion of a cloud, in a first approximation, Stokes' law can be used to calculate the steady state velocity of a cloud. Taking into consideration the fact that a cloud is not a solid, and therefore, there is internal circulation inside, theoretical calculations have shown that the numerical constant in Stokes' law should be changed from 3 to 2.5 (Lamb, 1945). This result hinges on the assumption that the viscosity of the cloud is the same as the viscosity of the air.

In fact, the viscosity of a cloud varies with the particle concentration of the cloud according to the expression (Brinkman, 1952, Roscoe, 1952)

$$\mu_c = \mu(1 - c')^{-k} \tag{2.47}$$

where c' is the concentration of the particles expressed as the volume of particles/total volume of suspension, and k is a constant equal to 2.5 for spheres. At low values of c', Eq. (2.47) simplifies:

$$\mu_c = \mu(1 + kc') \tag{2.48}$$

For larger values of the Reynolds number one could use the drag coefficient correlation (Fig. 2.2) and Eq. (2.21) or Eq. (2.22) to estimate the steady state velocity of the cloud. It is evident that as the cloud diameter is much greater than the diameter of a single particle, the friction force (resistance) acting on a cloud of particles at a given relative velocity is also much greater than the value for a single particle.

The terminal settling velocity in the region of validity of Stokes' law is directly proportional to the sphere diameter squared and to the effective density of the sphere [see Eq. (2.16)]. As the diameter of a cloud is usually great, one might expect a cloud to have a very high terminal settling velocity. The effective density of a cloud, however, is usually very close to that of the surrounding air, and therefore, the settling velocity is usually very low. In addition, updrafts hinder the settling of clouds.

Bounded Clouds

When there is a dusty gas in a duct or in a vessel, the cloud of particles is bounded by the walls of the equipment. In this case the frictional resistance encountered by individual particles is considered and is denoted by F_c. Under these conditions the frictional resistance encountered by the particle is greater than that predicted by Stokes' law. When there are no disorientations of the particles because of interparticle forces, the frictional resistance of a particle F_c has been expressed empirically by the equation (Lewis *et al.*, 1949; Richardson and Zaki, 1954)

$$F_c = F(1 - c')^{-4.65} \qquad (2.49)$$

where F is the frictional resistance given by Stokes' law, and c' is the concentration of particles (volume of particles/total volume). At low values of c' Eq. (2.42) simplifies:

$$F_c = F(1 + 4.65c') \qquad (2.50)$$

In cases where there are interparticle interactions, this relationship has been derived for low concentrations (Burgers, 1941, 1942; Hawlsley, 1950):

$$F_c = F(1 + 6.875c') \qquad (2.51)$$

The immediate consequence of Eqs. (2.49)–(2.51) is that a cloud of particles contained in a duct or a container will settle at a slower rate than if only an individual particle were settling, because according to these equations, the particles encounter a greater frictional resistance than an individual particle would under the same conditions. This phenomenon is often called "hindered settling" of particles. The proximity of many other particles interferes with the settling of each particle. Experimental measurements on hindered settling of particles have been performed (Richardson and Zaki, 1954) to values as high as $c' = 0.5$, where a 25-fold increase in the frictional resistance of the fluid could be observed.

2.6 Fluid Resistance to Particle Motion in the Proximity of Solid Boundaries

In the presence of solid surfaces, the frictional resistance of a fluid to particle motion becomes greater. This situation arises in filters of all kinds, where the filter fibers or granules are often very close to the particles passing through the filter.

The fluid resistance to particle motion near a solid boundary, F_W, is calculated by dividing the Stokes' law resistance F by the boundary correction factor K:

$$F_W = F/K \tag{2.52}$$

The value of K has been calculated theoretically for various geometries of the system as a function of the ratio d/l, for relatively small values of this ratio (l is the distance between the walls).

A good example of the magnitude of K is shown by the experimental results plotted in Fig. (2.4) (Francis, 1933). It can be seen from the curve that the increase in frictional resistance of the fluid becomes about fivefold

Figure 2.4 Experimental correction factor for fluid resistance for spheres moving along the axis of a cylinder (Francis, 1933).

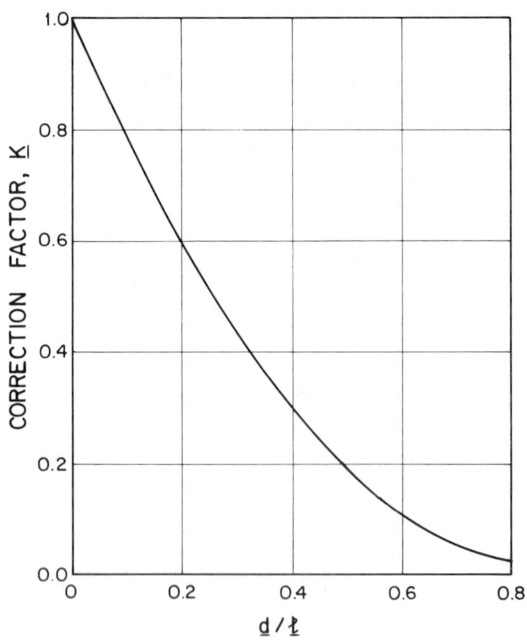

when the particle diameter is about $\frac{1}{2}$ of the cylinder diameter (l) in which the particle is moving.

2.7 Fluid Resistance to the Motion of Nonspherical Particles

The majority of particles encountered in practice are not spherical. The problem of predicting the fluid resistance to nonspherical particles can often be too difficult to solve. Theoretical predictions of the resistance are limited to relatively simple particle shapes such as ellipsoids, cylinders, and rectangular parallelepipeds.

Apart from the problems introduced by the large variety of possible different particle shapes, an additional difficulty is presented by the fact that nonspherical particles will encounter different resistance by the fluid, depending on the orientation in which they are traveling. Additional complication arises when certain particles are not settling vertically but along an inclined path. The most difficult and practically intractable cases are those of instabilities, when, for example, the particles wobble from side to side, spin, sideslip, or follow a spiral path. In general it can be recommended as a rule of thumb that the higher the symmetry of the particle and the more isometric the particle is, the more it can be expected to retain a stable orientation. For example, completely isometric particles such as cubes, tetrahedra, and octahedra can be most relied upon to retain their initial orientation.

The orientation of particles also depends on the value of the particle Reynolds number. For example, as the Reynolds number increases between 10 and 20, even tetrahedra and cubes tend to take up preferred orientations. At even higher Reynolds numbers, between 70 and 300, particle instability sets in even in the case of isometric particles.

The behavior of clouds of nonspherical particles has not been studied sufficiently to make a discussion possible. It can be generally assumed that orientational instabilities are less likely to occur when the particles are present in high concentrations.

The behavior of a nonspherical particle is usually referred to the behavior of a so-called equivalent sphere, that is, a sphere that has some property equal to that of the nonspherical particle under consideration. Some of the pertinent definitions are listed in Table 2.3 (21). The surface diameter of a non-spherical particle can be visualized by imagining that a thin surface layer of a particle is peeled off and made into a spherical shape. The volume diameter can be visualized by imagining that the nonspherical particle is made into a spherical shape. As of all shapes, the sphere has the smallest surface-to-volume ratio, so it follows that the surface diameter of a non-

Table 2.3 Equivalent Diameters and Shape Factors

Name	Symbol	Definition
Stokes (sedimentation) diameter	d_{st}	$[18\mu u_t/\rho_p g C(d_{st})]^{1/2}$, i.e., the diameter of a sphere of the same density as that of the particle which has the same terminal settling velocity as the particle ($F_p > F_s$)
Aerodynamic diameter	d_{Ae}	$[18\mu u_t/gC(d_{Ae})]^{1/2}$, i.e., the diameter of a unit density sphere with the same terminal settling velocity as the particle
Surface diameter	d_s	$(S/\pi)^{1/2}$, i.e., diameter of a sphere with the same external surface area as the particle
Volume diameter	d_v	$(6V/\pi)^{1/3}$, i.e., diameter of a sphere with the same volume as the particle
Area diameter	d_A	$(4A/\pi)^{1/2}$, i.e., diameter of a circle with the same area as the projected area of the particle
Drag diameter	d_e	Diameter of a sphere with the same resistance to motion as the particle, in a fluid of the same viscosity and at the same velocity $\left[F_p = F_s = \frac{3\pi\mu u_t d_e}{C(d_e)} = \frac{d_v^3\pi}{6}\rho_p g\right]$
Sphericity	$\Psi = (d_v/d_s)^2 \leq 1$	Ratio of surface area of a sphere with same volume as the particle to the actual surface area of the particle (S_s/S_p)
Circularity	$\chi \leq 1$	Ratio of the circumference of a circle having the same cross-sectional area as the irregular particle to the actual perimeter of the irregular particle's projected image ($P_c/P_p = d_A\pi/P_p$)
Dynamic shape factor	$X = d_e/d_v = (d_v/d_{st})^2$ ($C = 1$)	Ratio of terminal settling velocity of a sphere of the same mass and volume as the particle to the terminal settling velocity of the particle[a] $[(u_s)_t/u_t]$

[a] The terminal settling velocity, $(u_s)_t$, of a sphere of the same mass and volume as the particle is obtained from the steady state force balance [$C(d_v)$ = Cunningham correction factor of a sphere of diameter d_v]:

$$\frac{3\pi\mu(u_s)_t d_v}{C(d_v)} = \frac{d_v^3\pi}{6}\rho_p g, \qquad \therefore (u_s)_t = \frac{d_v^2\rho_p g C(d_v)}{18\mu}$$

The terminal settling velocity of the particle, u_t, may be obtained in terms of d_e and d_v [$C(d_e)$ = Cunningham correction factor of a sphere of diameter d_e]:

$$\frac{3\pi\mu u_t d_e}{C(d_e)} = \frac{d_v^3\pi}{6}\rho_p g, \qquad \therefore u_t = \frac{d_v^3\rho_p g C(d_e)}{d_e 18\mu}$$

spherical particle is always greater than its volume diameter. The square of the ratio volume diameter to surface diameter is called the sphericity of the nonspherical object, and this quantity is widely used to measure the deviation from spherical shape.

Probably the most important equivalent diameter in aerosol mechanics is the so-called drag diameter, meaning the diameter of a sphere for which the frictional fluid resistance F is the same as in the case of the motion of the nonspherical particle in question, *given that the fluid has the same viscosity and that the relative velocity between the fluid and the particle is the same in both cases.*

Calculated Drag Diameters of Spheroids

A spheroid is a body of revolution, meaning that it is generated by rotating an ellipse about its major or minor axis. (It *does not* mean that the body is rotating during its fall.) The particular axis chosen is called the "axis of revolution" of the spheroid. The various cases considered in these calculations are shown in Fig. 2.5 (Davies, 1947). The direction of fall is designated as a, and the direction perpendicular to it is b. In column (a) of Fig. 2.5, the spheroids are imagined moving along their respective axes of revolution, that is, axis a in this case. Note that

$$n \equiv a/b \tag{2.53}$$

When $n > 1$, a is the semimajor axis; for $n < 1$, b is the semimajor axis. The bottom figure represents a cigar-shaped spheroid falling with its nose pointing downward. The one above it is a sphere, and the third from the bottom is a spheroid shaped somewhat like a hamburger bun falling with its face down. The top figure in column (a) shows an elliptical plate or "coin" moving in a direction perpendicular to its own plane.

In column (b) of Fig. 2.5 the spheroids are imagined moving perpendicular to their respective axes of revolution, that is, axis b, in this case. Starting at the bottom, the first figure represents the cigar-shaped spheroid falling sideways, the second figure represents a sphere, and the third from the bottom represents the hamburger-bun-shaped object falling sideways. Finally, the top figure corresponds to the elliptical plate falling edgeways, in the direction of the a axis.

The results of the calculations have been plotted in Fig. 2.6(a) and (b). Curve i corresponds to the case where the spheroid is falling in the direction of its axis of revolution; curve ii corresponds to the case where the spheroid is falling at right angles to its axis of revolution; curve iii corresponds to the case of an elliptical plate moving sideways; and finally, the curve in Fig.

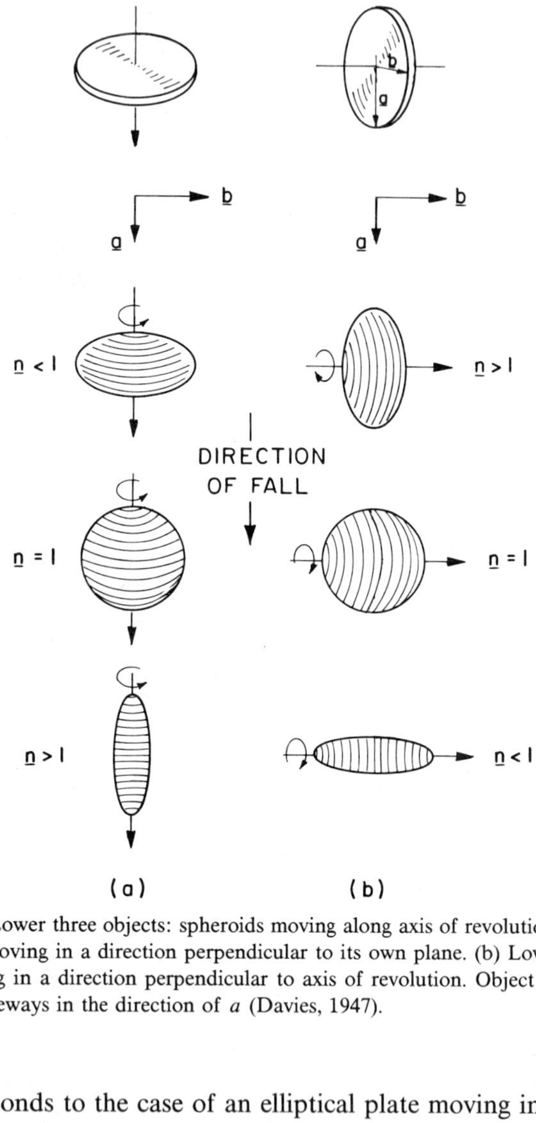

Figure 2.5 (a) Lower three objects: spheroids moving along axis of revolution. Object at top: elliptical plate moving in a direction perpendicular to its own plane. (b) Lower three objects: spheroids moving in a direction perpendicular to axis of revolution. Object at top: Elliptical plate moving sideways in the direction of a (Davies, 1947).

1.6(b) corresponds to the case of an elliptical plate moving in the direction perpendicular to its own plane.

The application of these diagrams is now explained. The appropriate curve is selected, and for the given value of n, the corresponding value of $d_e/2b$ is read from the graph. The size of the spheroid must be stated in the problem; hence, the magnitude of the axis b is known. Using b, the equivalent drag diameter d_e is calculated. When substituted into Stokes' law, d_e gives the frictional resistance encountered by the spheroid under the conditions specified in the problem.

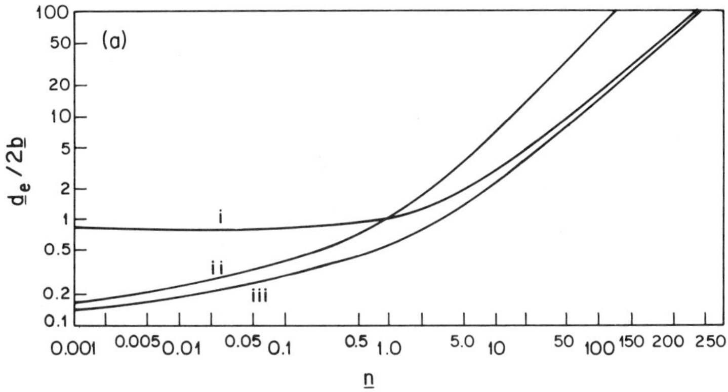

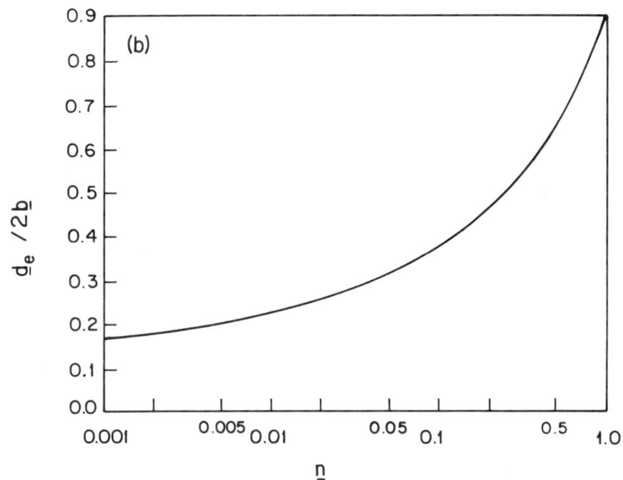

Figure 2.6 Equivalent diameters d_e for ellipsoids of revolution (Davies, 1947). (a) Moving in the direction of the a axis: (i) a = axis of rotation, $a = nb$; (ii) b = axis of rotation, $a = nb$; (iii) elliptical plate moving sideways in direction of a axis, $a = nb$. (b) Elliptical plate moving in the direction perpendicular to its own plane, $a = nb$.

Correlations for the Fluid Resistance to the Motion of Nonspherical Particles

Two correlations for nonspherical particles are shown in Figs. 2.7 (Heiss and Coull, 1952) and 2.8 (Pettyjohn and Christiansen, 1948).

On the left-hand side of Fig. 2.7, the curves are applicable for cylinders, spheroids, and rectangular parallelepipeds with the projected area in the direction of motion circular or equilateral. On the right-hand side of the same figure, the curves refer to cases where the projected area is not circular

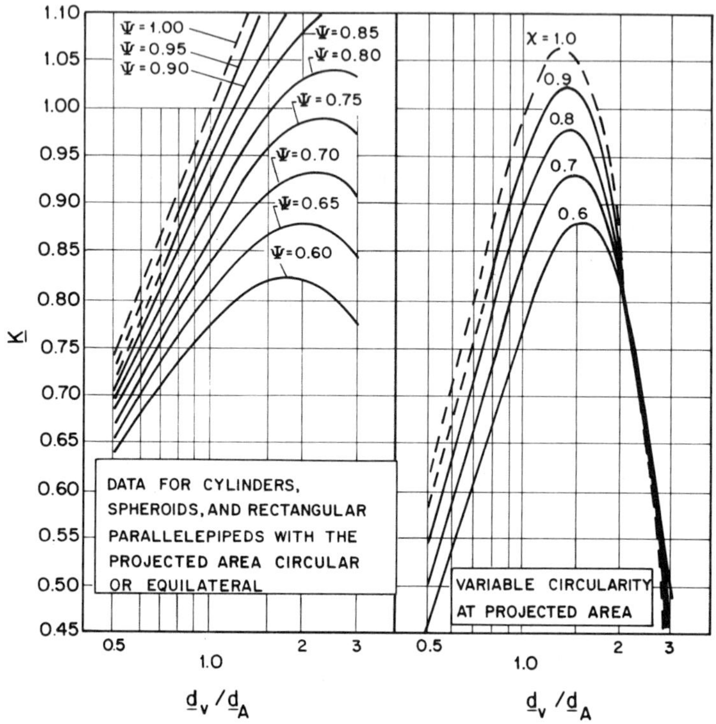

Figure 2.7 Correlation of settling factor K as a function of the sphericity Ψ and circularity χ for various ratios d_v/d_A for nonisometric particles in the viscous flow region (Heiss and Coull, 1952).

or equilateral. In both cases, the settling factor K is given as a function of the ratio volume diameter/area diameter. In the left-hand side of the figure, the parameter is the sphericity, whereas in the right-hand part, it is the circularity.

The particle velocity is obtained as now explained. First, calculate the settling velocity of a sphere of the equivalent volume diameter d_v, and then multiply this by the correction factor K.

For isometric particles, that is, cubes, octahedra, and tetrahedra, the correction factor K can also be calculated by the formula (Pettyjohn and Christiansen, 1948)

$$K = 0.843 \log_{10}(\psi/0.065) \tag{2.54}$$

The correlations shown in Fig. 2.7 are good only in the range of validity of Stokes' law. The correlation shown in Fig. 2.8 is limited to isometric particles; however, *it extends over a wide range of Reynolds numbers outside the validity of Stokes' law.* The parameter used is the sphericity of the

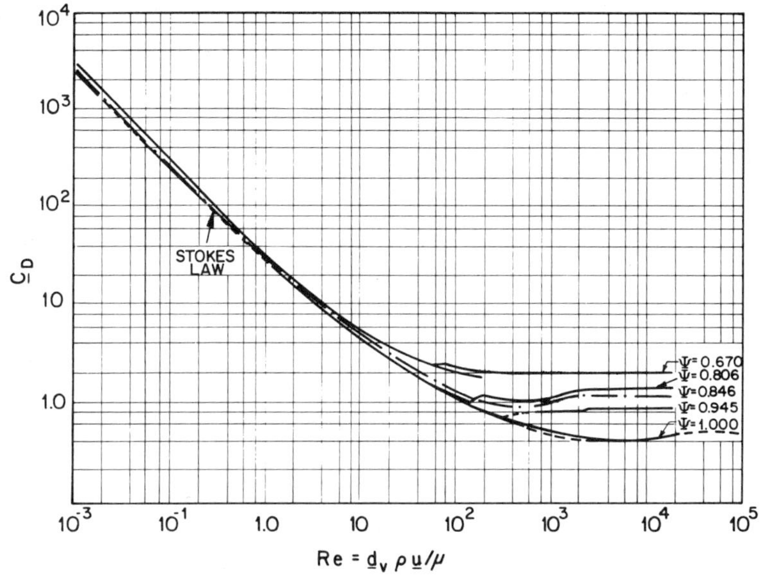

Figure 2.8 Correlation of drag coefficient C_D for isometric particles with different sphericity Ψ (Pettyjohn and Christiansen, 1948).

particle. The value of the drag coefficient read from this graph is used in exactly the same way as in the case of spherical particles. For Reynolds numbers greater than 2,000, the value of the drag coefficient can be calculated by the simple formula (Pettyjohn and Christiansen, 1948)

$$C_D = 5.31 - 4.88\psi \tag{2.55}$$

2.8 Effect of Particle Roughness

Just as in the case of pipe flow, surface roughness of particles influences the friction drag acting on particles. In general it can be said that the rougher the particle the greater the frictional resistance under otherwise identical conditions. No detailed studies of the effect of particle roughness on frictional resistance have been made so far.

2.9 Density of Agglomerates

One must be very careful about what particle density to use in the case of particles that are formed by agglomeration of smaller particles. In this case, the effective or bulk density of the particle agglomerate can be very different from the density of the pure solid of which the particles are made. Agglomerate densities have been back calculated from observed terminal settling velocities of agglomerates of known size.

EXAMPLE 2.1

Calculate the terminal settling velocity of a spherical particle with a 50-μm diameter and a density of 8×10^3 kg m^{-3} in still air of 1 atm pressure at 344°C ($\mu = 3.13 \times 10^{-5}$ Pa s).

SOLUTION

$$u_t = \frac{d^2(\rho_p - \rho)g}{18\mu} = \frac{(50)^2(10^{-12})\ \mathrm{m}^2\ (8 \times 10^3)\ \mathrm{kg\ m}^{-3}\ (9.81)\ \mathrm{m\ s}^{-2}}{(18)(3.13 \times 10^{-5})\ \mathrm{Pa\ s}}$$

$$= 0.348\ \mathrm{m\ s}^{-1}$$

Check whether this falls in the range of Stokes' law:

$$\mathrm{Re} = \frac{u_t \rho d}{\mu} = \frac{(0.348)\ \mathrm{m\ s}^{-1}\ (0.5)\ \mathrm{kg\ m}^{-3}\ (50 \times 10^{-6})\ \mathrm{m}}{(3.13 \times 10^{-5})\ \mathrm{Pa\ s}} = 0.28$$

This is low enough for Stokes' law to apply with only a very small error.

EXAMPLE 2.2

Calculate the distances traveled by a spherical particle with a density of 10^3 kg m^{-3} and a diameter of 40 μm, falling freely in still air of 1 atm pressure and 20°C temperature, before it reaches 99 and 99.9% of its terminal settling velocity. Assume that the initial velocity of the particle is equal to zero.

SOLUTION

According to the data in Table 3.1, the experimental settling velocity is 4.8×10^{-2} versus the value of 5.0×10^{-2} predicted by Stokes' law. This amounts to an error of only 4%. We are going to assume that Sotkes' law is valid.

We have these relationships:

$$u/u_t = 0.99$$

$$u = \tau g[1 - \exp(-t/\tau)]$$

$$u_t = \tau g$$

$$u/u_t = 1 - \exp(-t/\tau) = 0.99$$

giving $t = -\tau \ln(0.01)$.

By definition,

$$\tau = \frac{d^2\rho_p}{18\mu} = \frac{(16 \times 10^{-10})\ \mathrm{m}^2\ (10^3)\ \mathrm{kg\ m}^{-3}}{(18)(1.82 \times 10^{-5})\ \mathrm{Pa\ s}} = 4.88 \times 10^{-3}\ \mathrm{s}$$

giving $t = -(4.88)(10^{-3})\ln(0.01) = 2.25 \times 10^{-2}$ s.

Further, we have the relationship

$$x = \tau g\{t + \tau[\exp(-t/\tau) - 1]\}$$

$$\therefore x_{99} = (4.88 \times 10^{-3})(9.81)[2.25 \times 10^{-2} + (4.88 \times 10^{-3})(-0.99)]$$

$$= 8.46 \times 10^{-4} \text{ m}$$

Similarly, we can obtain the result for 99.9% of u_t:

$$x_{99.9} = (4.88 \times 10^{-3})(9.81)[3.37 \times 10^{-2} + (4.88 \times 10^{-3})(-0.999)]$$

$$= 1.38 \times 10^{-3} \text{ m}$$

EXAMPLE 2.3

Predict the terminal settling velocity of 1-mm-diameter water droplets falling in air at ambient conditions: temperature, 20°C; pressure, 101 kPa. Use the methods of von Kármán and of McCabe and Smith. Compare the predictions with the experimental results presented in Table 2.1.

SOLUTION

1. Method of von Kármán:

$$C_D \text{ Re}^2 = \frac{4\rho(\rho_p - \rho)d^3 g}{3\mu^2} = \frac{(4)(1.19)(10^3 - 1.19)(10^{-3})^3(9.81)}{(3)(1.82 \times 10^{-5})^2}$$

$$= 47{,}000$$

Hence, Eq. (2.22) applies:

$$\log \text{Re} = -1.29536 + 0.986 \log(C_D \text{ Re}^2) - 0.046677[\log(C_D \text{ Re}^2)]^2$$

$$+0.0011235[\log(C_D \text{ Re}^2)]^3 = 2.41$$

Thus, Re = 255, and

$$u_t = \frac{\text{Re}\,\mu}{d\rho} = \frac{(255)(1.82 \times 10^{-5})}{(10^{-3})(1.19)} = 3.90 \text{ m s}^{-1}$$

Experimental result from Table 2.2: $u_t = 3.82 \text{ m s}^{-1}$.

% Error = +2.1%

2. Method of McCabe and Smith:

$$K = d\left[\frac{\rho g(\rho_p - \rho)}{\mu^2}\right]^{1/3} = 10^{-3}\left[\frac{(9.81)(1.19)(10^3 - 1.19)}{(1.82 \times 10^{-5})^2}\right]^{1/3} = 32.8$$

Flow is in the transition region and

$$b_1 = 18.5, \qquad n = 0.6$$

$$\therefore u_t = \left[\frac{4d^{(n+1)}g(\rho_p - \rho)\rho^{(n-1)}}{3b_1\mu^n}\right]^{1/(2-n)} = \left[\frac{4d^{1.6}g(\rho_p - \rho)^{-0.4}}{(3)(18.5)\mu^{0.6}}\right]^{1/1.4}$$

$$= \left[\frac{(4)(10^{-3})^{1.6}(9.81)(10^3 - 1.19)(1.19)^{-0.4}}{(3)(18.5)(1.82 \times 10^{-5})^{0.6}}\right]^{1/1.4} = 4.14 \text{ m s}^{-1}$$

% Error = +8.2%

EXAMPLE 2.4

A spherical particle of density 6×10^3 kg/m^3, having a diameter of 0.5×10^{-3} m, is settling in air of density 1.2 kg/m^3 and viscosity of 1.8×10^{-5} Pa s. Calculate

1. the terminal settling velocity of the particle;
2. the time it takes for the particle to reach 99% of the terminal settling velocity if the particle started settling from rest; and
3. the distance covered by the particle in this time, using Eq. (2.40).
4. Check on your results by solving the equations of motion of the particle numerically, using Euler's method. Use a time increment of $h = 0.2$ s.

$$d = 0.5 \times 10^{-3} \text{ m} \qquad \rho = 1.28 \text{ kg m}^{-3}$$

$$\rho_p = 6 \times 10^3 \text{ kg m}^{-3} \qquad \mu = 1.8 \times 10^{-5} \text{ Pa s}$$

SOLUTION

1. Inspection shows Stokes' law does not apply. Use von Kármán's method:

$$C_D \text{Re}^2 = \frac{4\rho(\rho_p - \rho)d^3 g}{3\mu^2} = \frac{(4)(1.2)(6 \times 10^3 - 1.2)(0.5 \times 10^{-3})^3(9.81)}{(3)(1.8 \times 10^{-5})^2}$$

$$\therefore C_D \text{Re}^2 = 36.325 \times 10^3$$

$$\log \text{Re} = 1.29536 + 0.986 \log(C_D \text{Re}^2) - 0.046677[\log(C_D \text{Re}^2]^2$$

$$+0.0011235[\log(C_D \text{Re})^2]^3 = 2.33688$$

$$\therefore \text{Re} = 217.21 = \frac{u_t d\rho}{\mu}$$

$$\therefore u_t = \frac{\mu \text{Re}}{d\rho} \frac{(1.8 \times 10^{-5})(271.21)}{(0.5 \times 10^{-3})(1.2)} = 6.5163 \text{ m s}^{-1}$$

2. $u = 0.99u_t = 6.4512 \text{ m s}^{-1}$

$$t = \frac{1}{3}\frac{\rho_p d^2}{\sqrt{C_{DA}\rho}}\sqrt{\frac{\pi}{2G}} \ln \frac{(2/d)\sqrt{(2G/\pi C_{DA}\rho)} + u}{(2/d)\sqrt{(2G/\pi C_{DA}\rho)} - u}$$

Assume

$$C_{DA} = C_D(u_t) = \frac{36.326 \times 10^3}{(217.21)^2} = 0.76994$$

$$G = mg = \frac{d^3\pi}{6}(\rho_p - \rho)g = \frac{(0.5 \times 10^{-3})^3\pi}{6}(6 \times 10^{+3} - 1.2)(9.81)$$

$$= 3.8516 \times 10^{-6} N$$

$$\therefore t = \frac{1}{3}\frac{(6 \times 10^3)(0.5 \times 10^{-3})2}{\sqrt{(0.76994)(1.2)}}\sqrt{\frac{\pi}{(2)(2.8516 \times 10^{-6})}}$$

$$\times \ln\left[\frac{\frac{2}{0.5 \times 10^{-3}}\sqrt{[(2)(3.8516 \times 10^{-6})/\pi(0.76994)(1.2)]} + 6.4512}{\frac{2}{0.5 \times 10^{-3}}\sqrt{[(2)(3.8516 \times 10^{-6})/\pi(0.76994)(1.2)]} - 6.4512}\right]$$

$$= 1.75875 \text{ s}$$

3. $$x = \frac{4}{3}\frac{\rho_p d}{\rho C_{DA}}\ln\cosh\left[\frac{3\sqrt{2C_{DA}G}}{2d^2\rho_p\pi^{1/2}}t\right]$$

$$\therefore x = \frac{4}{3}\frac{(6 \times 10^3)(0.5 \times 10^{-3})}{(1.2)(0.76994)}$$

$$\times \ln\cosh\frac{\left[\frac{3(2)(0.76994)(1.2)(3.8516 \times 10^{-6})}{2(0.5 \times 10^{-3})^2(6 \times 10^3)\pi^{1/2}} \times 1.7587\right]}{2.64708}$$

$$= 8.4809 \text{ m}$$

4. a. $x(t_n) = x(t_{n-1}) + h \cdot u(t_{n-1})$

b. $u(t_n) = u(t_{n-1}) + h \cdot a(t_{n-1})$

c. $$\text{Re}(t_n) = \frac{u(t_n)\,d\rho}{\mu} = \frac{u(t_n)(0.5 \times 10^{-3})(1.2)}{(1.8 \times 10^{-5})} = 33.333u(t_n)$$

d. $$C_D(t_n) = \frac{24}{\text{Re}(t_n)} + \frac{4}{[\text{Re}(t_n)]^{1/3}} \qquad \text{[Eq. (2.8)]}$$

e. $$C_D(t_n)[\text{Re}(t_n)]^2 = \frac{4d^3\rho\rho_p[g - a(t_n)]}{3\mu^2}$$

$$\therefore g - \frac{3\mu^2 C_D(t_n)[\text{Re}(t_n)]^2}{4d^3\rho\rho_p} = a(t_n)$$

$$\therefore a(t_n) = 9.81 - \frac{(3)(1.8 \times 10^{-5})^2 C_D(t_n)[\text{Re}(t_n)]^2}{(4)(0.5 \times 10^{-3})^3(1.2)(6 \times 10^3)}$$

$$\therefore a(t_n) = 9.81 - 270.00 \times 10^{-6} C_D(t_n)[\text{Re}(t_n)]^2$$

Initial conditions

$$x(t_0) = u(t_0) = 0$$

$$a(t_0) = 9.81 \text{ m s}^{-2}$$

Use

$$h = 0.2 \text{ s}$$

n	t_n (s)	$x(t_n)$ (m)	$u(t_n)$ (m/s)	$\text{Re}(t_n)$	$C_D(t_n)$	$C_D(t_n)[\text{Re}(t_n)]^2$	$a(t_n)$ (m/s^2)
0	0		0	0	0	0	9.81
1	0.2	0	1.9620	63.399	1.3714	5,512.1	8.3217
2	0.4	0.3924	3.6263	120.88	1.0075	14,722	5.8351
3	0.6	1.1177	4.7933	159.78	0.88736	22,654	3.6934
4	0.8	2.0763	5.5320	184.40	0.83291	28,322	2.1631
5	1.0	3.1827	5.9646	198.82	0.80606	31,863	1.2070
6	1.2	4.3776	6.2060	206.86	0.79236	33,906	0.65535
7	1.4	5.6168	6.3371	211.23	0.78526	35,037	0.35002
8	1.6	6.8843	6.4071	213.57	0.78156	35,645	0.18483
9	1.8	8.1657	6.4441	214.80	0.77963		

Interpolate for $t = 1.7587$,

$\therefore\ x = 7.9011$ m

This is less than 8.4809 m by -6.8%. Check for C_D at u_t:

$$C_D = \frac{24}{217.21} + \frac{4}{(217.21)^{1/3}} = 0.77592 \neq 0.76994$$

There is a small discrepancy between the two different relationships used to relate Re to C_D. Is this partly responsible for the discrepancy in x? Repeating the calculation, using the "ln cosh" formula with $C_D = 0.77952$ and $t = 1.7587$ s, we get $x = 8.3127$ m. Had we used $C_D = 0.77963$, we would have come even closer to $x = 7.9011$ m.

The fairly good agreement between the results obtained by Eq. (2.40) and by Euler's method is because of a number of circumstances, but mainly, because of the fact that C_D approaches its terminal value fairly closely quite early in the fall.

EXAMPLE 2.5

Calculate the terminal settling velocity for an oblate spheroid

1. falling in the direction of its axis of revolution and
2. falling in the direction perpendicular to its axis of revolution.

Then compare the results with the terminal settling velocity of a sphere of the same density and the same mass as the spheroid, where

$$\rho_p = 2 \times 10^3 \text{ kg/m}^3$$

$$\rho = 1.2 \text{ kg/m}^3$$

$$\mu = 1.82 \times 10^{-5} \text{ Pa s}$$

the length of the semimajor axis = 10 μm, and the length of semiminor axis = (10/3) μm.

SOLUTION

Recall that an oblate spheroid is one that is formed by rotating an ellipse about its minor axis. This results in a hamburger-bun-shaped object.

1. When the oblate spheroid is falling in the direction of its axis of revolution the minor axis points in the a direction. Thus,

$$n \equiv a/b = \text{length of semiminor axis/length of semimajor axis}$$

$$n = (10/3)/10 = 1/3$$

Using Fig. 2.6(a), curve (i),

$$d_e/2b \cong 0.9 \qquad \text{and} \qquad d_e = (2)(10)(0.9) = 18\ \mu\text{m}$$

because b, the length of the semi-axis perpendicular to the direction of fall, is in this case the semimajor axis. The volume of an oblate spheroid is given by

$$V = 4/3\pi \times (\text{length of semimajor axis})^2 \times (\text{length of semiminor axis})$$

or

$$V = 4/3\pi b^2 a$$

(in the notation used in this part of the problem). From the definition of equivalent drag diameter, d_e, there follows:

$$3\pi\mu d_e u_t = \frac{d_v^3\pi}{6}(\rho_p - \rho)g$$

$$\therefore\ u_t = \frac{4b^2a(\rho_p - \rho)g}{9\mu d_e} = \frac{(4)(10)^2(10/3)(2 \times 10^3 - 1.2)(9.81)(10^{-12})}{(9)(1.82 \times 10^{-5})(18)}$$

$$= 8.87 \times 10^{-3} \text{ m s}^{-1}$$

Checking with the Reynolds number,

$$\text{Re} = \frac{d_e u_t \rho}{\mu} = \frac{(18 \times 10^{-6})(8.87 \times 10^{-3})(1.2)}{1.82 \times 10^{-5}} = 0.011$$

Hence, Stokes' law applies.

2. In this case the direction of fall, that is, the a direction is the one in which the major axis points. Thus,

$$n \equiv a/b = \text{length of semimajor axis/length of semiminor axis}$$

$$n = 10/(10/3) = 3$$

Using Fig. 2.6(a), curve (ii),

$$d_e/2b \cong 2.5$$

where b is the length of the semiminor axis. Therefore,

$$d_e = (2.5)(2)(10/3) = 16.667\,\mu\text{m}$$

The terminal velocity is given by

$$u_t = \frac{4a^2b(\rho_p - \rho)g}{9\mu d_e} \qquad \text{(in the notation used in this part of the problem)}$$

$$u_t = \frac{(4)(10)^2(10/3)(2 \times 10^3 - 1.2)(9.81)(10^{-12})}{(9)(1.82 \times 10^{-5})(16.667)}$$

$$= 9.58 \times 10^{-3}\ \text{m s}^{-1}$$

The volume diameter, d_v, is calculated as follows. The volume of the spheroid $= \frac{4}{3}\pi a^2 b$. Here, a = length of semimajor axis, and b = length of semiminor axis. The volume of the sphere $= \pi d_v^3/6$. Therefore,

$$\pi d_v^3/6 = 4/3\pi a^2 b$$

$$\therefore d_v = (8a^2b)^{1/3} = [(8)(10)^2(10/3)]^{1/3} = 13.867\,\mu\text{m}$$

The terminal settling velocity of the sphere, using Stokes' law, is

$$u_t = \frac{d_v^2 g(\rho_p - \rho)}{18\mu}$$

$$\therefore u_t = \frac{(13.867 \times 10^{-6})^2(9.81)(2 \times 10^3 - 1.2)}{(18)(1.82 \times 10^{-5})} = 1.15 \times 10^{-2}\ \text{m s}^{-1}$$

Summary of results

Sphere	$u_t = 1.15 \times 10^{-2}$ m s^{-1}	
Spheroid falling in the direction of its axis of revolution	$u_t = 0.887 \times 10^{-2}$ m s^{-1}	% Difference = −22.9%
Spheroid falling perpendicular to axis of revolution	$u_t = 0.958 \times 10^{-2}$ m s^{-1}	% Difference = −16.7%

PROBLEMS

1. A spherical particle with a 10-μm diameter and a density of 1.92 g/cm^3 is placed on a horizontal screen. Air is blown through the screen vertically at 20°C temperature and 1 atm pressure. Calculate

   a. the air velocity required to just lift the particle,

   b. the particle Reynolds number at this condition,

   c. the drag force exerted by the air on the particle, and

   d. the drag coefficient.

2. A mixture of quartz and hornblende is to be separated by sedimentation in air. The mixture is present in a layer 2 cm thick before it is dropped into the sedimentation tube. The mean particle diameters are 500 μm for the quartz and 600 μm for the hornblende. The densities are 2.6 g/cm^3 for the quartz and 3.5 g/cm^3 for the hornblende. Assume spherical particles.

   a. Calculate the distance(s) over which the particles must fall before the mixture is completely separated.

   b. Calculate the time it takes to cover that distance.

   HINT: Assume that all particles have the same diameter that is equal to the mean diameter and that there is no interaction between the particles while they fall.

3. A stack of a steel mill at a height 120 m is emitting iron-oxide fumes into a 5 km/h wind. The particle size ranges from 0.1 to 20 μm. Estimate the distance downwind that can be expected to be free of iron-oxide deposit. Use 5.5 g/cm^3 for the density of iron oxide and assume spherical particles.

4. Calculate the terminal settling velocity of a spherical cloud of particles in air where the diameter of the cloud, d_c, is 1 m

$$d_p = 1\ \mu\text{m}$$

$$\rho_p = 1\ \text{g/cm}^3$$

$$\rho = 1.195 \times 10^{-3}\ \text{g/cm}^3$$

$$\mu = 1.84 \times 10^{-4}\ \text{g/cm s}$$

and the particulate loading in the cloud is 0.1 grains/ft^3.

5. A new cereal process is being planned. The new cereal is spherical in shape, and it is prepared by spraying a warm liquid horizontally into the top of a 32-ft column. The average drop diameter of the liquid leaving the nozzle is 5,000 μm. This diameter remains relatively constant during the solidification process while the spheres dry in the column. The time required for drying in the hot air (100°C) is approximately 1.2 s.

   The viscosity and the density of the hot moist air are estimated at 2.1×10^{-5} Pa s and 0.8 kg m^{-3}, respectively. The average specific gravity of the cereal during the drying process is estimated at 0.8.

   a. Calculate the vertical distance covered by the cereal in 1.2 s while falling in the column, by taking into account the period of acceleration, in order to ascertain that the column is tall enough for complete drying to be achieved. State your conclusions.

   b. Repeat the calculation, but this time assume that the cereal falls at its terminal settling velocity over the entires 1.2 s. State your conclusions.

6. Compare the free fall in still, ambient air of three spherical particles 1, 2, and 3, such that $d_1 = 0.1$ μm, $d_2 = 1$ μm, and $d_3 = 10$ μm. All particles have specific gravity equal to 5.

   a. Calculate the time it will take each particle to reach 99% of its terminal settling velocity.

   b. Calculate the distance covered by each particle until reaching 99% of its terminal settling velocity.

   c. Calculate the terminal settling velocity of each particle.

7. Estimate the terminal settling velocity in air at 20°C and 1 atm of an ellipsoid (oblate spheroid) moving along its axis of revolution, with

   a = axis of revolution

   $a = 0.2b, \qquad b = 2\mu$

   The density of the spheroid = 5 gm/cm^3. Do the calculation by two methods: (1) using Fig. 2.6(a), and (2) using the left-hand part of Fig. 2.7. Compare the results.

   HINT: Volume of spheroid = $4/3\pi b^2 a$

   Surface of spheroid $= 2\pi b^2 + \pi \dfrac{a^2}{\varepsilon} \ln \dfrac{1+\varepsilon}{1-\varepsilon}$

   where

$$\varepsilon = \sqrt{1 - \left(\frac{a}{b}\right)^2}$$

8. A spherical particle of specific gravity 1.5 has a diameter = 2 mm. The particle is allowed to fall freely in air at 20°C and 760 mm Hg pressure.
   a. Estimate the terminal settling velocity assuming that Stokes' law is valid.
   b. Estimate the terminal settling velocity by using Eq. (2.22) in your text. Check whether this equation is applicable and comment.
   c. Based on your result, estimate the distance, x, the particle has covered in the first 3 s of its free fall.
   d. Calculate the distance, x, the particle covered in the first 3 s of its free fall using both methods demonstrated in Example 2.4.

References

Allen, H. S. (1900). Motion of a sphere in a viscous fluid. *Philos. Mag.* **50**, 323–338, 519–536.

Bird, R. B., Stewart, W. E., and Lightfoot, E. N. (1960). "Transport Phenomena." Wiley, New York.

Brinkman, H. C. (1952). The viscosity of concentrated suspensions and solutions. *J. Phys. Chem.* **20**, 571.

Burgers, J. M. (1941). Effect of concentration on the speed of sedimentation of suspensions, especially suspensions of spherical particles. *Proc. K. Ned. Akad. Wet.* **44**, 1045–1051, 1177–1184 (1941); **45**, 9, 126 (1942).

Cunningham, E. (1910). Terminal velocity of spheres falling in a fluid. *Proc. Soc. London A* **83**, 357–365.

Davies, C. N. (1945). Definitive equations for the fluid resistance of spheres. *Proc. Phys. Soc. London* **57**, 259–270.

Davies, C. N. (1947). The sedimentation of small suspended particles. *London Symp. Particle Size Analysis, Inst. Chem. Eng.*, p. 12–26.

Francis, A. W. (1933). Wall effect in falling ball method for viscosity. *Physics* **4**, 403–460.

Fuchs, N. A. (1964). *In* "Mechanics of Aerosols" (C. N. Davies, ed.), revised Ed., trans. Pergamon, Oxford.

Happel, J., and Brenner, H. (1965). "Low Reynolds Number Hydrodynamics." Prentice-Hall, Englewood Cliffs, New Jersey.

Hawlsley, P. G. W. (1950). Some aspects of fluid flow. *Inst. Phys. Conf.* p. 114.

Heiss, J. F., and Coull, J. (1952). The effect of orientation and shape on the settling velocity of nonisometric particles in a viscous medium. *Chem. Eng. Prog.* **48**, 133–140.

Klyachko, L. (1934). Otopl. i Ventil. (4) (quoted by Fuchs, N. A.). *In* "Mechanics of Aerosols" (C. N. Davis, ed.), p. 33, revised Ed., trans. Pergamon, Oxford (1964).

Lamb, H. (1945). "Hydrodynamics." Dover, New York.

Langmuir, I., and Blodgett, K. B. (1946). Ann. Air Force Tech. Rep., 5418.

Lewis, W. K., Gilliland, E. R., and Bauer, W. C. (1949). Characteristics of fluidized particles. *Ind. Eng. Chem.* **41**, 1104–1117.

Lunnon, R. G. (1928). Fluid resistance to moving spheres. *Proc. R. Soc. London A* **118**, 680–694.

McCabe, W. L., and Smith, J. C. (1976). "Unit Operations of Chemical Engineering," 3rd Ed., p. 154. McGraw-Hill, New York.

Pettyjohn, E. S., and Christiansen, E. B. (1948). Effect of particle shape on free-settling rates of isometric particles. *Chem. Eng. Prog.* **44**, 157–172.

Richardson, J. F., and Zaki, W. N. (1954). Sedimentation and fluidization. *Trans. Inst. Chem. Eng.* **32**, 35–53.

Roscoe, R. (1952). Viscosity of suspensions of rigid spheres. *Br. J. Appl. Phys.* **3**, 267–297.

Schiller, L., and Naumann, A. (1973). Ueber die grundlegenden Berechnungen bei der Schwerkraftaufbereitung. *VDI-Z* **77**, 318–320.

Schlichting, H. (1962). "Boundary Layer Theory." Pergamon, Oxford.

Strauss, W. (1975). "Industrial Gas Cleaning," 2nd Ed. Pergamon, Oxford.

Torobin, L. B., and Gauvin, W. H. Fundamental aspects of solid-gas flow. *Can. J. Chem. Eng.* **37**, 129–141, 167, 224–236 (1959); **38**, 142–153, 189–200 (1960); **39**, 113–120 (1961).

3

Gravity, Momentum, and Centrifugal Separators

3.1 Introduction

In this chapter, some of the laws of fluid resistance to particle motion, studied in the preceding chapter, are utilized in the design of certain types of particle separators which are widely used in industry. These may be conveniently divided into three categories: gravity, momentum, and centrifugal separators.

In gravity separators, the particles are permitted to settle to the bottom of a chamber under the influence of gravity. In momentum separators, the gas stream in a settling chamber is diverted by baffles and, as a result of inertia, the particles will impinge against the baffle or the walls of the chamber. In centrifugal separators, centrifugal force is used to make the particles impinge against the walls of a cylindrical vessel. The deposit formed on the walls falls to the bottom of the equipment where it is continually removed.

All of these methods use gravitational or inertial forces, which make the particles impinge against solid surfaces contained in the equipment. Once the particles have settled out of the gas stream, they tend to agglomerate and also to adhere to the solid surfaces to some extent and, hence, are not picked up by the gas stream as easily as the individual particles. Nevertheless, reentrainment is always a problem to contend with, and this should be minimized by designing the equipment so that a certain "critical" gas velocity is not exceeded in those parts containing deposited dust.

3.2 Gravity Settlers

A gravity settler contains a chamber in which the gas velocity is reduced in order to give the particles enough time to settle out by the action of gravity

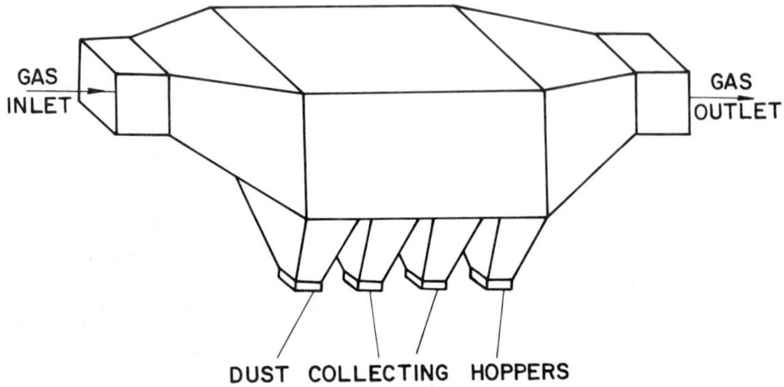

Figure 3.1 Horizontal flow settling chamber with square cross section. [Reprinted with permission from Strauss (1975).]

and also to minimize reentrainment. The settling chamber is usually horizontal and rectangular with an inlet at one end and an outlet at the other, as shown in Fig. 3.1.

Settling chambers usually give 100% removal efficiencies for particles that are about 100 μm in diameter or greater. For particles smaller than about 50 μm, the removal efficiency is usually very poor. The efficiency also depends on particle density and shape, and on the details of the design of the settling chamber. Example 3.1 indicates that with long residence times 100% removal efficiencies may be obtained for much smaller particles. But, in reality, small particles tend to be mixed by the turbulent eddies present in the settler and, therefore, settle much slower than predicted by the existing methods, which do not take turbulent mixing into account. With increasing particle size, the particles follow less and less the turbulent eddies of the gas, and therefore, the predictions become more accurate for larger particles.

These assumptions are usually made in settling chamber design.

1. Plug velocity profile of gas flow exists.
2. There is uniform inlet particle concentration.
3. Particles start settling at their terminal settling velocity.
4. The Cunningham correction factor, C, is equal to one.
5. There is no reentrainment.
6. Stokes' law applies.
7. The particles are spherical.
8. Interparticle interaction is neglected.

When the dusty gas stream enters the settling chamber, some particles are at a higher elevation, whereas others are at a lower level. Evidently, those particles that enter that chamber at the top will take the longest to settle out.

If H is the height of the settling chamber, and particles of a certain diameter d settle a distance h during the residence time t_r, then the ratio h/H represents the fraction of these particles that will be collected. Everything else being equal, the greater the particle diameter, the bigger is h/H. A curve of h/H versus particle size for a given material is the fractional or grade efficiency curve of the settling chamber for that particular material.

The residence time, t_r, is given by the equation

$$t_r = \frac{L}{v} = \frac{V}{Q} = \frac{LBH}{Q} \tag{3.1}$$

where L is the length of the chamber, v the average gas velocity in the chamber, B the breadth of the chamber, Q the volume flow rate of the gas, and V the volume of the chamber. It is important to note that as $\eta = h/H = (t_r u_t)/H = (LBHu_t)/QH = (LBu_t/Q)$, for a given Q, η is the same value for different values of L and B, provided that the product LB is the same and $v = Q/(HB) < v_{\text{pickup}}$.

It is customary to use the mean falling speed, u_m, of the particle, which is defined as

$$u_m = h/t_r \tag{3.2}$$

For the purpose of grade efficiency curve calculations, the terminal settling velocity u_t is a good approximation of the mean falling speed u_m, under most conditions of practical interest. The difference between u_t and u_m becomes noticeable only for particles of such great size that they would travel in the time t_r a distance much greater than the height of the settling chamber. As these particles are evidently collected with 100% efficiency, there is no need to calculate their collection efficiency, and the assumption $u_m = u_t$ is not required for these particles.

In the range of validity of Stokes' law Eq. (2.16) is used to calculate u_t. Outside the range of validity of Stokes' law, either Eqs. (2.21) and (2.22) or, in a good approximation, Eqs. (2.23) and (2.24) may be used for the same purpose. It is usually found, however, that those particles that are too big for Stokes' law to apply are collected with 100% efficiency. Table 3.1 shows that at ambient conditions Stokes' law applies to water droplets of diameters up to about 14 μm.

For any given set of conditions, it is possible to calculate the minimum particle size $d_{\min}$ that will be theoretically wholly retained within the chamber by calculating the value of d corresponding to the known values of H (height of chamber), t_r (residence time), ρ_p, ρ, and μ. The calculation may be done, depending on the conditions, by using Eq. (2.16), Eq. (2.21),

Table 3.1 Settling Velocity of Spherical Particles in Air[a, b]

Particle diameter (μm)	Experimental ($m\ s^{-1}$)	Calculated from Stokes' law ($m\ s^{-1}$)
0.1	8.7×10^{-7}	8.71×10^{-7}
0.2	2.3×10^{-6}	2.27×10^{-6}
0.4	6.8×10^{-6}	6.85×10^{-6}
1.0	3.5×10^{-5}	3.49×10^{-5}
2	1.19×10^{-4}	1.19×10^{-4}
4	5.0×10^{-4}	5.00×10^{-4}
10	3.06×10^{-3}	3.06×10^{-3}
20	1.2×10^{-2}	1.2×10^{-2}
40	4.8×10^{-2}	5.0×10^{-2}
100	2.46×10^{-1}	3.0×10^{-1}
400	1.57	4.83
1000	3.82	30.50
= 1 mm		

[a]Density, 1000 kg m^{-3}; temperature, 20°C; pressure, 100 kPa.
[b]Reprinted with permission from Strauss (1975).

Eq. (2.22), or Eqs. (2.23) and (2.24). Often, Eq. (2.40) can be solved explicitly for d, and then it is very convenient for the purpose of this calculation.

Pickup velocity in a settling chamber is the velocity at which reentrainment starts. Typical pickup velocities of various materials are listed in Table 3.2.

Table 3.2 Pickup Velocities of Various Materials[a]

Material	Density (kg $m^{-3} \times 10^3$)	Median size (μm)	Pickup velocity ($m\ s^{-1}$)
Aluminium chips	2.72	335	4.3
Asbestos	2.20	261	5.2
Nonferrous foundry dust	3.02	117	5.7
Lead oxide	8.26	14.7	7.6
Limestone	2.78	71	6.4
Starch	1.27	64	1.8
Steel shot	6.85	96	4.6
Wood chips	1.18	1370	3.9
Wood sawdust	—	1400	6.8

[a]From Baliff *et al.* (1948).

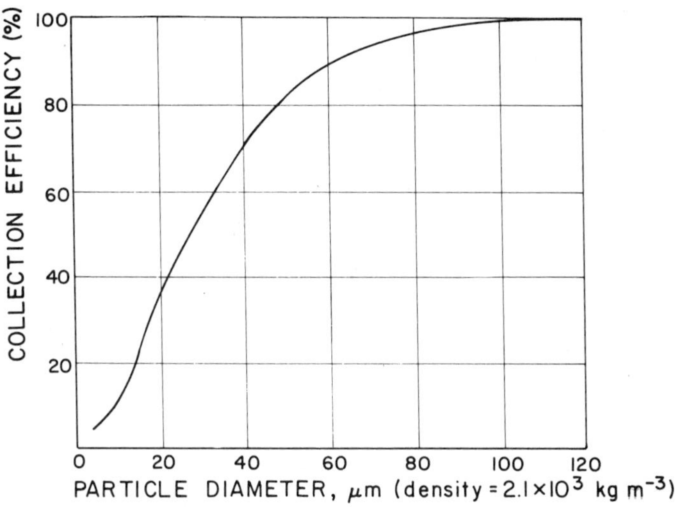

Figure 3.2 Fractional efficiency curve for a vertical flow settling chamber (Baliff *et al.*, 1948).

Settling chambers are indispensable for the collection of coarser grit. Their major advantage is the low cost and the dry disposal of collected materials. Because of the low velocities prevailing in the settling chambers, an additional advantage is that there are no problems with abrasive materials and virtually no pressure drop to overcome.

Settling chambers often handle vast volumes of gas flow. In a typical installation of a sinter plant, the volume handled is 2.3×10^6 m^3/h. With an inlet dust loading of 1 g/m^3, this is equivalent to 2.3 tons of solids per hour. With such large volumes, the low power consumption associated with the low pressure loss of the unit is of great economic importance. A typical grade efficiency curve showing collection efficiency versus particle diameter for particles of density 2.1×10^3 kg/m^3 is shown in Fig. 3.2.

3.3 Momentum Separators

The efficiency of a simple settling chamber can be improved and the space requirements can be reduced by giving the particles a downward momentum in addition to the gravity effect. Simple baffle type collectors are shown in Fig. 3.3. More sophisticated and modern development of momentum separators are shown in Fig. 3.4. The gases flow down a cylindrical duct

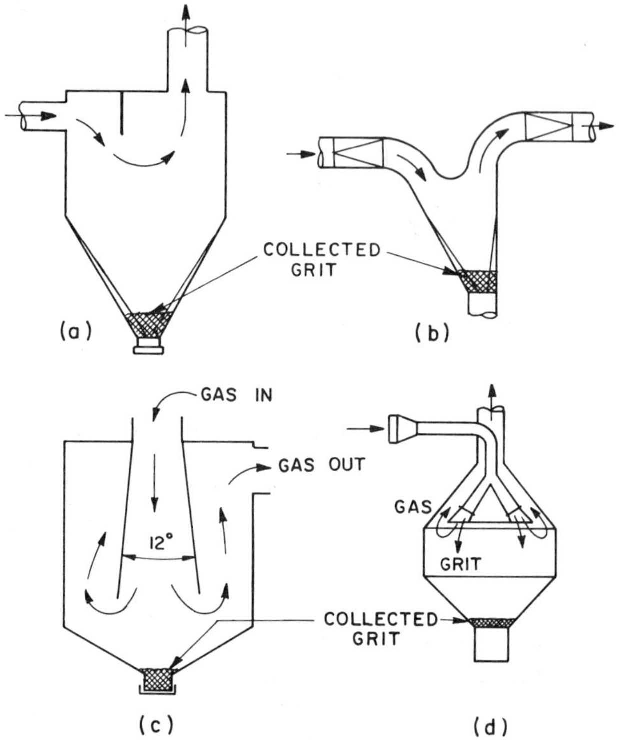

Figure 3.3 Baffle-type collectors where particles are slightly accelerated toward the hopper: (a) simple baffle type (Alden, 1959); (b) rounded inertial trap (Alden, 1959); (c) downward facing entry (Jennings, 1950); (d) accelerated settling chamber with deepened hopper (Procket, 1930).

and then change direction by 180° to pass up an annular duct surrounding the inlet, while the dust is deposited in a hopper. In one case, additional rotating air is blown in at an annulus with a velocity three times that of the axial velocity in the main gas stream. In the second, less effective design, no secondary air is used. The fractional efficiency curves indicate that with elaborate design the effectiveness of momentum separators can be extended to particle sizes as small as 30 μm.

EXAMPLE 3.1

The installation of a gravity settler to help remove fly ash particles (specific gravity = 2.31) from an air stream at ambient conditions has been proposed. The inlet dust loading is 21 grains/ft^3, and the volumetric flow rate

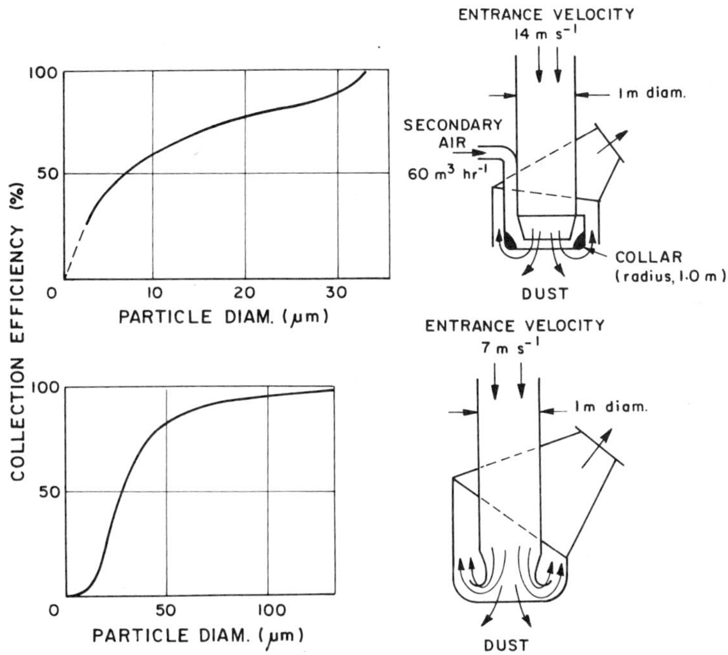

Figure 3.4 Modern momentum separators (designs and fractional efficiency curves) (Muschelknautz, 1967).

is 4,680 ACFM (actual cubic feet per minute). The inlet size distribution of the fly ash is given in this table.

Size range (μm)	Mass (%)
0–10	2
10–20	2
20–30	6
30–40	19
40–60	20
60–80	22
80–100	18
100–150	11

The dimensions of the unit are 10 ft wide × 20 ft high × 45 ft long.

1. Calculate an average collection efficiency over each size range. Make the usual assumptions. Repeat the calculations assuming that the particles start settling with zero velocity.

2. Estimate the overall collection efficiency of the unit.
3. Calculate the outlet dust loading, the daily mass of dust collected, and the daily mass of dust discharged from the unit.

SOLUTION

1. Let us first calculate the average velocity, v, in the chamber:

$$v = \frac{4{,}680}{10 \times 20 \times 60} \times 0.305 = 0.199 \text{ m s}^{-1}$$

This value is one to two orders of magnitudes less than the usual pickup velocities. The residence time is

$$t_r = \frac{LBH}{Q} = \frac{(10)(20)(45) \text{ ft}^3}{(4{,}680/60) \text{ ft}^3/\text{s}} = 115 \text{ s}$$

(note that this is an unusually high value). This unit is not used to capacity as it could easily handle ten times the gas flow specified in this problem.

$$K = d\left(\frac{g\rho\rho_p}{\mu^2}\right)^{1/3} = d\left[\frac{(9.81)\text{ m/s}^2\ (2.31 \times 10^3)\text{ kg/m}^3\ (1.2)\text{ kg/m}^3}{1.8^2 \times 10^{-10}\text{ Pa}^2\text{ s}^2}\right]^{1/3}$$

$$K = 43{,}783d \ (d \text{ in m}) \approx 4.38 \times 10^{-2}d \ (d \text{ in } \mu\text{m})$$

In Stokes' law range,

$$u_t = \frac{gd^2\rho_p}{18\mu} = \frac{(9.81)\text{ m/s}^2\ (2.31 \times 10^3)\text{ kg/m}^3}{(18)(1.8 \times 10^{-5})\text{ Pa s}}d^2$$

At first we let $u_m = u_t$:

$$u_m \text{ (m/s)} \approx 7.00 \times 10^7 d^2 \ (d \text{ in m})$$

$$h \text{ (m)} = u_m \text{ (m/s) } 115 \text{ s}; \qquad \eta = h/H = h\text{ (m)}/6.1 \text{ m}$$

Size range (μm)	$\bar{d}$ (μm)	K	$u_t = u_m$ (m/s)	h (m)	η
0–10	5	0.219	1.75×10^{-3}	2.01×10^{-1}	3.3×10^{-2}
10–20	15	0.66	1.58×10^{-2}	1.82	2.98×10^{-1}
20–30	25	1.09	4.38×10^{-2}	5.04	8.26×10^{-1}
30–40	35	1.53	8.58×10^{-2}	9.87	1.00
40–60	50	2.19			1.00
60–80	70	3.06			1.00
80–100	90	3.94			1.00
100–150	125	5.47			1.00

Assume now that the particles start settling from rest. In Stokes' law range:

$$h = \tau g[t_r + \tau(e^{-t_r/\tau} - 1)]$$

where τ = relaxation time = $d^2\rho_p/18\mu = u_t/g$. For $\bar{d} = 5\ \mu$m,

$$\tau = \frac{1.75 \times 10^{-3}}{9.81} = 1.78 \times 10^{-4}\ \text{s}$$

$$e^{-t_r/\tau} = e^{-115/1.78\times 10^{-4}} \approx 0$$

and

$$t_r - \tau = 115\ \text{s} - 1.78 \times 10^{-4}\ \text{s} \approx 115\ \text{s}$$

$$\eta = 3.3 \times 10^{-2}$$

For $\bar{d} = 70\ \mu$m,

$$\tau = \frac{3.43 \times 10^{-1}}{9.81} = 3.5 \times 10^{-2}\ \text{s}$$

$$e^{-t_r/\tau} = e^{-115/3.5\times 10^{-2}\ \text{s}} \approx 0$$

and

$$t_r - \tau = 115\ \text{s} - 3.5 \times 10^{-2}\ \text{s} \approx 115\ \text{s}$$

$$\eta = 1.00$$

Thus, the efficiencies are the same as before. Hence, the conclusion can be reached that the assumption that the particles start settling at a velocity u_t is a good one.

A word of caution is in order regarding the predicted grade efficiency curve. Some of the assumptions made in the calculations, in particular the tacit assumption of absence of turbulence implied by using Stokes' law, result in higher predicted removal efficiencies than observed in reality. Let us calculate the "duct" Reynolds number of the chamber:

$$(\text{Re})_{\text{chamber}} = \frac{vD_H\rho}{\mu}$$

$$= \frac{[4{,}680/(10 \times 20 \times 60)][4(10 \times 20)/2(10 + 20)](1.2)(0.305)^2}{1.85 \times 10^{-5}}$$

$$\approx 32{,}000$$

Evidently, the flow in the chamber is quite turbulent notwithstanding the very low gas velocity. Hence, turbulent mixing can be expected to interfere with the settling of the particles, and the more so the smaller the particle size.

2. $\bar{\eta} = 0.02 \times 0.033 + 0.02 \times 0.298 + 0.06 \times 0.825 + 0.19 + 0.20$
$+ 0.22 + 0.18 + 0.11 = 0.956$

3. $L_i = 21$ grain/ft^3
$L_f = (1 - 0.956)21 = 0.92$ grain/ft^3

Mass collected:

$$(21)\ \text{grain/ft}^3\ (4{,}680)\ \text{ft}^3/\text{min}\ (60 \times 24)\ \text{min/day}$$

$$\times (0.0648)\ \text{g/grain}\ (10^{-3})\ \text{kg/g}\ (0.956) = 9.17 \times 0.956$$

$$= 8.77 \times 10^3\ \text{kg/day}$$

Mass discharged:

$$9.17(1 - 0.956) = 403\ \text{kg/day}$$

EXAMPLE 2.2

Design a gravity settler to remove all particles greater than 75 μm (specific gravity = 1.91) from a dust-laden gas stream at ambient conditions. The volumetric flow rate is 100 ACFS (actual cubic feet per second). Submit your design. The maximum allowable throughput velocity is 10 ft/s. Make the usual assumptions.

SOLUTION

$$K = d\left(\frac{g\rho\rho_p}{\mu^2}\right)^{1/3} = 75 \times 10^{-6}\left(\frac{9.81 \times 1.2 \times 1.91 \times 10^3}{1.82^2 \times 10^{-10}}\right)^{1/3} = 3.06 < 3.3$$

Hence, Stokes' law applies:

$$u_t = \frac{75^2 \times 10^{-12} \times 1.91 \times 10^3 \times 9.81}{18 \times 1.82 \times 10^{-5}} = 0.322\ \text{m/s}$$

The residence time is $t_r = L/v = V/Q = LBH/Q$; for 100% removal, $t_r = H/u_t$.

u_t, v, Q	known
H, L, B	unknown
$L/v = H/u_t$ and $Q = vBH$	independent relationships

For the most economical design, set $L = B$. Thus,

$$H/u_t = \frac{L}{v} \quad \text{and} \quad Q = vLH$$

$$\therefore\ H/u_t = \frac{L}{Q/LH} = \frac{L^2H}{Q}$$

$$\therefore\ L = B = \sqrt{Q/u_t} = \sqrt{100 \times 2.82 \times 10^2/0.322} = 2.96\ \text{m}$$

$$H = \frac{L}{v}u_t = \frac{2.96}{3.045} \times 0.322 = 0.313\ \text{m}$$

But H should be equal to 1 m, at the very least, for cleaning purposes; v may be less than 10 ft/s. So we let $H = 1$ m and calculate the corresponding value of v:

$$v = \frac{Q}{BH} = \frac{100 \times 2.82 \times 10^{-2}}{2.96 \times 1} = 0.953 \text{ m/s}$$

This design results in a residence time

$$t_r = L/v = 2.96/0.953 = 3.1 \text{ s} = H/u_t$$

Hence, the theoretical removal efficiency of the unit is unchanged.

3.4 Cyclone Collectors

A cyclone collector is a device without moving parts in which the velocity of an inlet gas stream is transformed into a confined vortex from which centrifugal forces tend to drive the suspended particles against the wall of the cyclone body. Cyclones are among the most widely used dust collectors. They are relatively simple and inexpensive to make, are operated at moderate pressure losses, and can give long and trouble free service. They are mostly used for dusts of particle size greater than 5 μm in diameter, that is, lying in the size range between coarse grit, usually removed in settling chambers, and fine dust, removed by means of electrostatic precipitators, fabric filters, and wet scrubbers. Cyclone collectors are used in the collection and recycling of catalyst particles; they are also used as a step in the manufacture of freeze-dried coffee and flour, and as a precleaner before electrostatic precipitators in power generating stations using stokers.

The necessary parts of a cyclone are a gas inlet, a cylindrical body, an axial outlet for cleaned gas, and a dust discharge opening. These main types of cyclones are in use (see Fig. 3.5) (Caplan, 1977):

1. The common or conventional reverse flow cyclone with tangential inlet and axial dust discharge
2. The cyclone with tangential inlet and peripheral dust discharge
3. The cyclone with axial inlet through swirl vanes and axial dust discharge
4. The cyclone with axial inlet through swirl vanes and peripheral dust discharge

The common cyclone is by far the most widely used type, and discussion here is limited to this type. The properties of the vortex in the common cyclone are as follows. The gas entering the tangential inlet near the top of the cylindrical body creates a vortex or spiral flow downward between the walls of the gas outlet and the body of the cyclone (Fig. 3.6). This vortex, called the main vortex, extends downward to near the bottom of the cone.

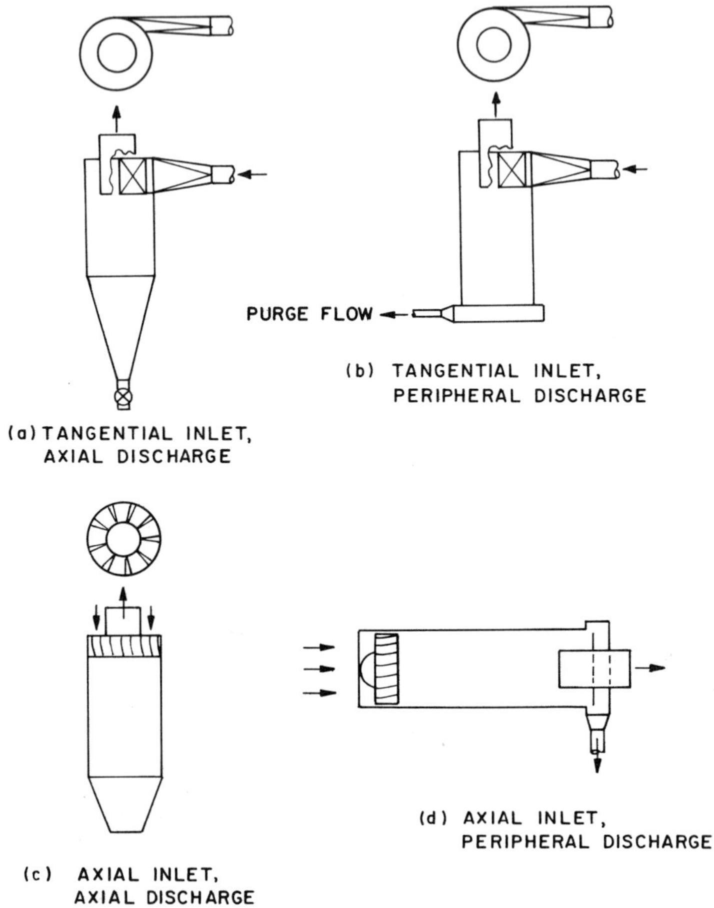

Figure 3.5 Types of cyclones in common use. [Reprinted with permission from Strauss (1975).]

Inside the main vortex, a secondary or inner vortex core is travelling upward to the gas outlet.

The tangential velocity of gas, v_T, increases with decreasing radial distance measured from the axis of the cylindrical body to a maximum value at some intermediate radial distance. In this region of increasing tangential velocity of the gas, v_T can be expressed by the formula

$$v_T = v_{T_p}(R/r)^{\gamma} \tag{3.3}$$

Here, v_{T_p} is the tangential velocity at the perimeter ($r = R$), r is the radial distance from the axis of the cylindrical body, and γ is a dimensionless

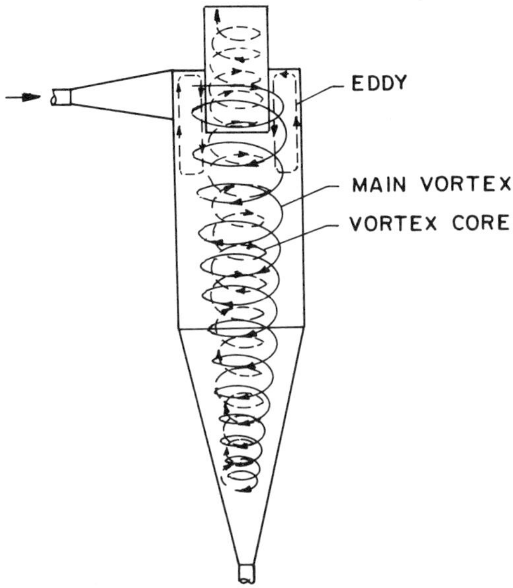

Figure 3.6 Vortex and eddy flows in a cyclone. [Reprinted with permission from Strauss (1975).]

vortex exponent, the value of which would be equal to one in the case of an ideal gas. The values for real gases are between 0.5 and 1, depending on the radius of the cyclone body and the gas temperature.

The inner vortex core is generally smaller in diameter than the gas outlet. The radius of the inner vortex core is usually between 0.2 and 0.4 times the radius of the gas outlet; the radius of maximum tangential velocity is usually from 0.4 to 0.8 times the radius of the gas outlet.

In the annular space between the cyclone body and the gas outlet, near the top of the body where the tangential inlet gas enters, the tangential gas velocity increases uniformly from the body wall to the outlet wall and there is generally a downward vortex flow. Because the radius of the outlet is greater than the radius of maximum velocity, the gas does not attain the maximum velocity in the annulus. In addition, there is upward gas flow along the body wall near the top of the cylinder. This upward flow, known as an eddy, carries gas and dust particles up along the body wall, inward across the top and downward along the gas outlet wall. From this point, dust particles are lost into the gas outlet, resulting in a lowering of cyclone efficiency.

Most particles, however, are separated from the gas by centrifugal force, which drives the particles toward the cyclone walls across the stream lines of the gas flow. The centrifugal force imparted to the particle is expressed as

$$G = mv_T^2/r \tag{3.4}$$

where G is the centrifugal force, m the mass of a particle, and v_T the particle tangential velocity.

By assuming that the tangential particle velocity is equal to the tangential gas velocity, this formula is obtained from Eqs. (3.3) and (3.4) for G (Caplan, 1977):

$$G = \beta \rho_p d^3 v_{Tp}^2 R^{2\gamma}/r^{(2\gamma+1)} \tag{3.5}$$

Here, β is a dimensionless shape factor.

3.5 Methods of Cyclone Design

A great deal of work has been done by various workers to predict the effectiveness of common cyclones. Predictions based on theoretical models have not been found suitable for design purposes. The design methods in use are empirical or semiempirical, notwithstanding the great ingenuity used in developing them. In cyclone design in general, it is recommended not to deviate very far from the construction of cyclones that have been actually built and on which experimental performance evaluations have been carried out. In cyclone design, the efficiency and pressure loss are predicted. Most predictive methods predict the entire grade efficiency curve, but some methods predict only the so-called "cut" of the cyclone, which is the diameter of particles (d_{50}) collected with 50% collection efficiency. The concept of cut diameter is used generally in operations in which particles are removed from a gas stream. In the next two sections, some of the more important predictive methods of cyclone efficiency and pressure loss are described.

Cyclone Collection Efficiency

Stairmand's Method

The idea underlying this method, limited to the prediction of cut diameters, is that particles spinning in a ring at the point of maximum tangential velocity v_{Tmax} have a 50% chance of being collected, or passing into the exit pipe. Hence, the diameter of such particles that spin in a ring at this radial distance from the center line corresponds to the cut diameter d_{50}. (The treatment does not pay attention to particles of the same diameter, d_{50}, that are not at this radial distance from the center line or, alternatively, it gives no reason why all such particles should be at this radial position.) It is assumed that these particles spin around, because the centrifugal force acting on them is exactly counterbalanced by the friction drag exerted on them by the radial inward drift of the gas. Therefore, for particles spinning around at the radial distance of maximum tangential velocity, the force balance can be written as

$$G = F \tag{3.6}$$

There is no unanimity regarding the radial distance from the axis of the cyclone body where the tangential velocity is maximum. Stairmand (1951) has suggested that this occurs at a radial distance $D_e/4$, ter Linden (1949) claimed this to occur at $D_e/3$, whereas according to Barth (1956), it occurs at $D_e/2$. (The various symbols referring to the main measurements of a cyclone, as used here, are explained in Fig. 3.7.) The same relationship is obtained between v_{Tmax} and d_{50} regardless at what radial distance the tangential velocity is assumed to be maximum. Let us denote this by $r_{\max}$. Thus,

$$G = mv_{\mathrm{Tmax}}^2/r_{\max} \tag{3.7}$$

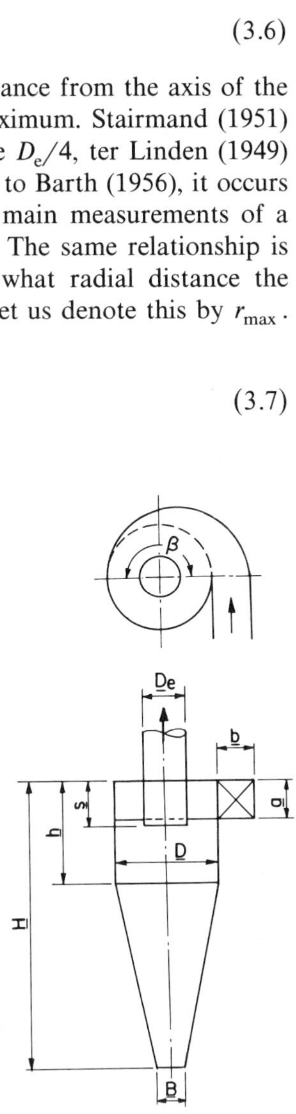

Figure 3.7 Conventional cyclone showing dimensions used in calculations. [Reprinted with permission from Strauss (1975).]

In the range of validity of Stokes' law, the friction drag is given by the expression

$$F = 3\pi\mu d\bar{u}_r \tag{3.8}$$

where $\bar{u}_r$ is the relative velocity of the particles with respect to the gas, which is equal to the average radial drift velocity of the gas, because, by assumption, the radial velocity of the particles is equal to zero:

$$\bar{u}_r = \frac{Q}{2\pi r_{max}(H - S)} \tag{3.9}$$

After substituting Eq. (3.9) into Eq. (3.8) and equating the expression thus obtained to Eq. (3.7), we see that r_{max} cancels out and we obtain

$$m v_{Tmax}^2 = \frac{3\mu d_{50} Q}{2(H - S)} \tag{3.10}$$

Assuming $r_{max} = D_e/4$, and that $v_{T_p} = v_i$ and $n = 0.5$ in Eq. (3.3), Stairmand obtained

$$v_{Tmax}^2(D_e/4) = v_i^2(D/2) \tag{3.11}$$

where v_i is inlet gas velocity. Equation (3.11) is used to eliminate v_{Tmax} from Eq. (3.10). Assuming that the particles are spherical, we obtain, by combining Eqs. (3.10) and (3.11), the expression for the "cut" of the cyclone:

$$d_{50} = \frac{3}{v_i \phi} \sqrt{\frac{\mu Q D_e}{2\pi\rho_p (H - S) D}} \tag{3.12a}$$

The empirical correction factor ϕ in Eq. (3.12) is called the friction loss factor. According to Stairmand (1949), ϕ can be calculated by the equation

$$\phi = \frac{-\sqrt{D_e/2(D - b)} + \sqrt{[D_e/2(D - b)] + (4\Gamma A/ab)}}{2\Gamma A/ab} \tag{3.13}$$

or, alternatively, it can be read from the graph given in Fig. 3.8. (Here, Γ is a friction constant, 0.005 for gas cyclones; and A is the surface area of the cyclone exposed to gases.)

Scale-Up Relationships of Stairmand

Stairmand and Kelsey (1955) determined the fractional efficiency curves for two types of cyclone, one being a high-efficiency, medium-throughout type, and the other, a medium-efficiency, high-throughput model. Both cyclone types are shown in Fig. 3.9. The measured fractional efficiency curves are shown in Fig. 3.10. The measurements were obtained for 200-mm diameter cyclones, dust density of 2,000 kg/m^3, in air at 20°C, and with an entrance

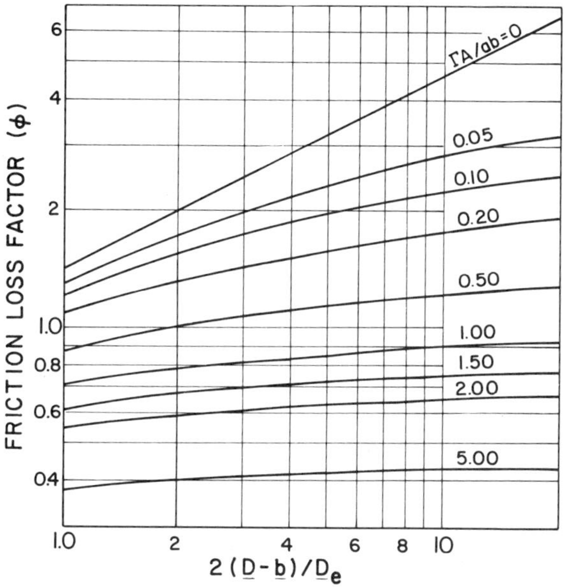

Figure 3.8 Curves for friction loss factor ϕ against entrance radius ratio for different cyclones (Stairmand, 1949).

velocity of approximately 15 m/s. It can be seen from these figures that the cyclone with smaller exit pipe diameter and smaller inlet duct cross section is more efficient than the cyclone where both the inlet and exit duct cross sections are greater and where the cylindrical body of the cyclone is "squat" because of the wide wraparound inlet used.

The efficiency curves given in Fig. 3.10 can be used to predict the efficiencies of the two cyclones under different operating conditions and for different types of dust, and also for cyclones of the same relative proportions, but of different size than the two models used by Stairmand and Kelsey.

The construction of the new fractional efficiency curve is accomplished by using the scale-up relationship which follows from a generalization of Eq. (3.12) to any collection efficiency. It gives the particle size of the new (1) dust that will be collected with the same efficiency as a particular selected size of the test dust (2):

$$d_1 = d_2\sqrt{\frac{\rho_{p2}Q_2\mu_1D_1^3}{\rho_{p1}Q_1\mu_2D_2^3}} \tag{3.12b}$$

where ρ_{p2}, Q_2, μ_2, and D_2 are the density of the test dust, the volume flow

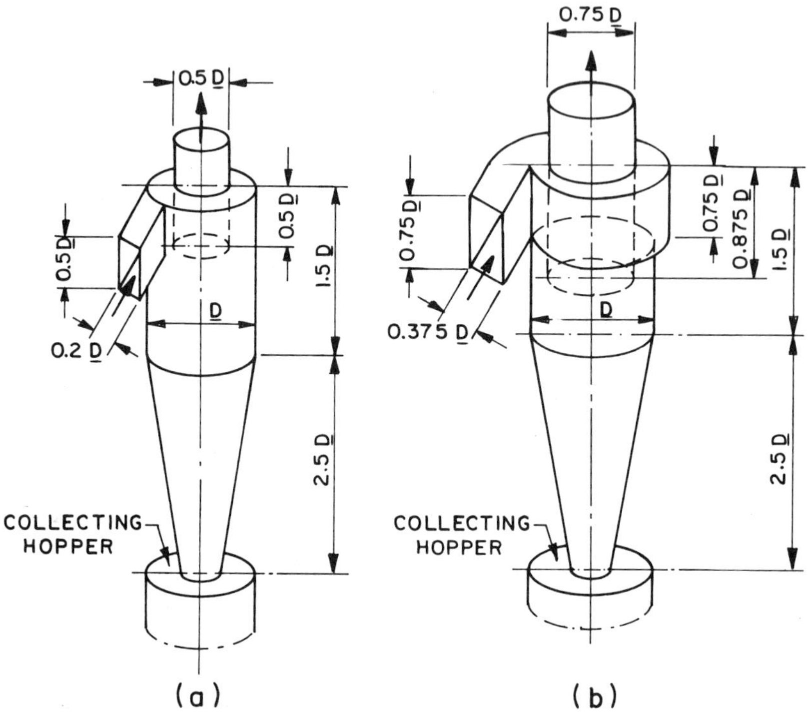

Figure 3.9 Standard cyclone designs (Stairmand). (a) High efficiency, medium throughput pattern. Normal flow rate = 5,450 diam^2 m^3 h^{-1}. (b) Medium efficiency, high throughput pattern. Normal flow rate = 16,470 diam^2 m^3 h^{-1}. Entrance velocity at these flows is approximately 15 m s^{-1} in both types (Stairmand and Kelsey, 1955).

in the test, the gas viscosity in the test, and the test cyclone diameter, respectively; and ρ_{p1}, Q_1, μ_1, and D_1 are the corresponding new quantities. It is noted that the relationship $v_i \propto Q/D^2$ is used when obtaining Eq. (3.12b) from Eq. (3.12a).

Barth's Method

The method developed by Barth (1956) to calculate the fractional efficiency of a cyclone is based on an idea similar to Stairmand's, that is, at the radial distance of maximum tangential velocity of the gas, the spinning particles have an even chance of being recovered in the cyclone or carried out by the gas stream. This method serves to predict not only the cut diameter of the cyclone, but the entire grade efficiency curve.

Barth introduced the quantity:

$$u_t^* = \bar{u}_r/n \tag{3.14}$$

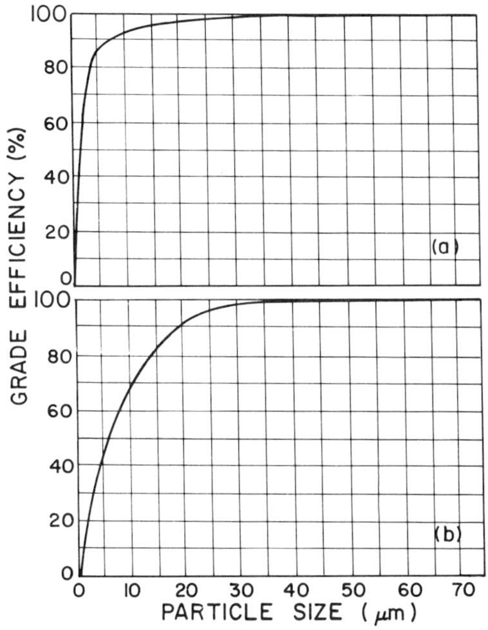

Figure 3.10 Fractional efficiency curves for 200-mm-diam cyclones (inlet velocity approximately 15 m s^{-1}; solid density 2000 kg m^{-3} in air at 20°C. (a) High-efficiency cyclone, Fig. 3.9(a); (b) high throughput cyclone, Fig. 3.9(b) (Stairmand and Kelsey, 1955).

where

$$\bar{u}_r = \frac{Q}{2\pi r_{max}(H - S)} \tag{3.15}$$

and

$$n = \frac{v_{Tmax}^2}{r_{max}\, g} \tag{3.16}$$

giving

$$u_t^* = \frac{Qg}{2\pi(H - S) v_{Tmax}^2} \tag{3.17}$$

The physical meaning of u_t^* is the average inward radial drift velocity of the gas at a radial distance r_{max} from the axis of the cylindrical body of the cyclone, divided by n, the number of g's at that distance. (g is the gravitational acceleration constant.) In other words, u_t^* is equal to the average radial drift velocity of the gas per g at the distance r_{max} from the

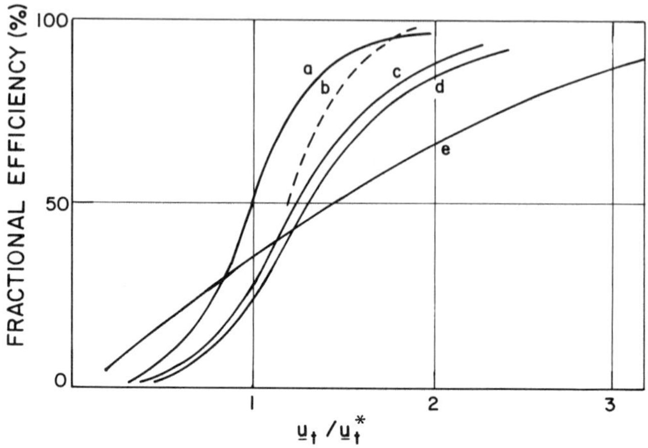

Figure 3.11 Curves for fractional efficiency calculations (Barth's method) (Muschelknautz and Brunner, 1967): a, experimental curve (Barth); b, hydrocyclone ($D = 77$ mm); c, cyclone with tangential entry ($D = 400$ mm); d, cyclone with tangential entry ($D = 300$ mm); e, axial entry cyclone ($D = 200$ mm).

axis of symmetry of the cyclone. By assumption, some particles have zero radial velocity at this distance from the axis, relative to the cyclone. As u_t is the terminal settling velocity of particles (under the influence of one g), the following relationship must hold for particles such that spin at r_{max}:

$$u_t = u_t^* \tag{3.18}$$

By the same assumption, used also by Stairmand, Eq. (3.18) is the condition defining the "cut" of the cyclone (see Fig. 3.11, curve a).

It has been assumed that particles for which the terminal settling velocity is greater than u_t^* will be recovered with an efficiency greater than 50%. These particles have a diameter greater than the cut of the cyclone. Conversely, particles for which the terminal settling velocity is less than u_t^* will be recovered with a fractional efficiency smaller than 50%. Using the model consistently, that is, if particles of diameter d_{50} are assumed to spin around with zero radial velocity, it should follow that all larger particles are collected with 100% efficiency and all smaller particles with 0% efficiency. This conclusion, however, is at odds with experience. The relationship shown in Fig. 3.11 was determined by Barth experimentally, and curve a in the figure was calculated on the basis of his measurements, using Eqs. (3.17) and (2.16). The other curves in Fig. 3.11 were obtained subsequently by Muschelknautz and Brunner (1967) on cyclones where the values v_{Tmax}/v_i were measured. Some relevant geometrical and physical data on these cyclones (shown in Fig. 3.11) are listed in Tables 3.3 and 3.4. It is evident

Table 3.3 Cyclone Proportions for the Fractional Efficiencies Shown in Fig. 3.11[a]

Curve	Cyclone type[b]	$\frac{a \times b}{\pi/4 D_e^2}$	$\frac{D}{D_e}$	$\frac{2b}{D}$	$\frac{2H}{D_e}$	$\frac{v_{Tmax}}{v_i}$
b	(D)	0.96	3	0.3	10	2.3
c	E	0.44	4	0.17	7.5	4.3
d	D	0.90	3	0.3	10	2.8
e	A	(0.75)	2	(0.36)	10	2.0

[a]From Muschelknautz and Brunner (1967).
[b]The cyclone types are those shown in Fig. 3.12.

from Fig. 3.11 that for different types of cyclones different fractional efficiency versus u_t/u_t^* curves apply. One can also check the validity of Eq. (3.8) by using the measured values of v_{Tmax}/v_i, listed in Tables 3.3 and 3.4. The various types of cyclones that were investigated are shown diagrammatically in Fig. 3.12.

The entire fractional efficiency curve of a cyclone can be estimated as explained here. For the given volume flow Q, the value of u_t^* is calculated by Eq. (3.17), using a measured or assumed value for v_{Tmax}. Then u_t is calculated by Eq. (2.16) over the entire particle size distribution of the dust. The ratio u_t/u_t^* is formed for each u_t, and the corresponding efficiency is read from the appropriate curve in Fig. 3.11. Finally, the efficiencies are plotted versus the corresponding particle diameters.

Table 3.4 Cyclone Proportions for Different Cyclone Types and Their Experimental Pressure Loss Factors ζ and Loss Numbers[a]

Cyclone type[b]	A	B	C	D	E
D/D_e	2	3	3.5	3	4
$(a \times b)/(\pi D_e^2/4)$	2.7	0.9	1[c]	0.9	0.44
$2H/D_e$	12.5	23.5	14	14.5	11
$2b/D$	(0.40)	0.27	0.57	0.27	0.17
α	0.76	0.76	0.71	0.75	0.93
Γ, dimensionless wall friction coefficient	0.005	0.005	0.005	0.005	0.010 (0.007)
v_{Tmax}/v_i	0.74	2.6	2.9	3	4.3
ε	12.5	2.73	2.5	2.45	2.0
ζ	6.9	18.5	21	22	37

[a]From Muschelkrautz and Brunner (1967).
[b]The cyclone types are those shown in Fig. 3.12.
[c]Here it is b^2/D_e^2.

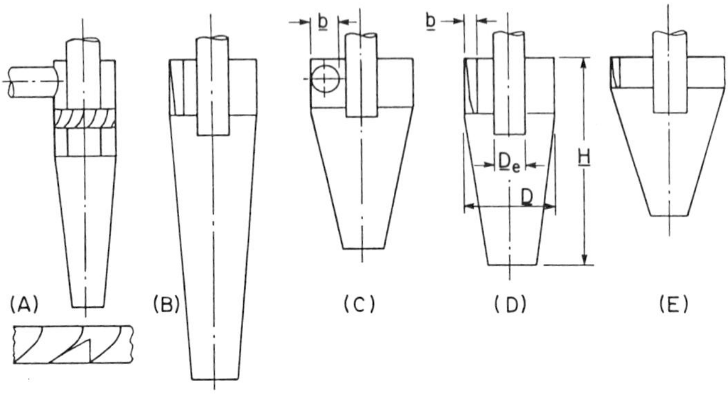

Figure 3.12 Cyclone models used by Muschelknautz and Brunner (1967). Their relative dimensions are given in Tables 3.3 and 3.4.

The cut diameter predicted by Barth's method will be identical with the value obtained by using Eq. (3.12) provided that the same value of v_{Tmax}^2 and $\phi = 1$ are used. Indeed, it can be readily shown that Eq. (3.12) follows from Eqs. (3.18), (3.17), (3.11), and (2.16). Using the measured ratios $v_{\mathrm{Tmax}}/v_{\mathrm{i}}$, instead of Eq. (3.11), will, in general, result in different values for d_{50}.

Lapple's Method

This semi-empirical approach (Lapple, 1951) of cyclone efficiency calculation utilizes this design equation for the cut diameter:

$$d_{50} = 3\sqrt{\frac{\mu b}{2\pi N_{\mathrm{t}} v_{\mathrm{i}}(\rho_{\mathrm{p}} - \rho)}} \tag{3.19}$$

where N_{t} is the effective number of turns the gas stream makes in the cyclone; N_{t} is given in a first approximation by the formula

$$N_{\mathrm{t}} = \frac{(V/Q)v_{\mathrm{i}}}{\pi D} \tag{3.20}$$

where V, the "useful" volume of the cyclone is

$$V = \frac{\pi}{4}\left[\left(\frac{H-h}{D-B}\right)\left(\frac{D^3 - B^3}{3}\right) + D^2 h - D_{\mathrm{e}}^2 S\right] \tag{3.21}$$

The simplified physical picture used is that those particles that travel in the residence time in the outer vortex a distance equal to $\frac{1}{2}$ of the width b of the inlet duct are collected with 50% efficiency. The radial velocity of these particles is estimated as $u_{\mathrm{t}}' = d_{50}^2 \rho_{\mathrm{p}} v_{\mathrm{i}}^2 / 18\mu(D/2)$. The residence time is

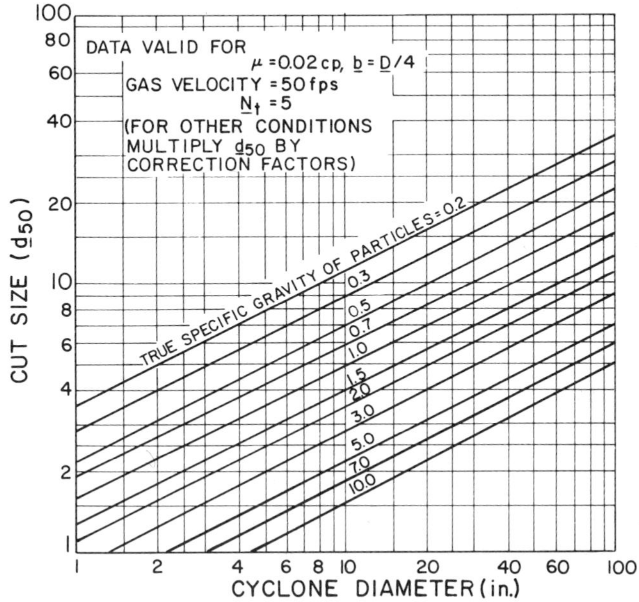

Figure 3.13 Cut size in microns for cyclones of conventional type (Lapple, 1951).

$t_r = N_t \pi D / v_i$. Inserting these relations into $b/2 = u_t' t_r$, Eq. (3.19) is obtained. A similar argument might be applied to particles that are collected with $10, 20, \ldots, 90, 100\%$ efficiency, resulting in the entire predicted grade efficiency curve. Unfortunately, the predictions do not agree with the experiment. For cyclones with the geometrical proportions $b = D/4$, $D_e = D/2$, $a = D/2$, $h = 2D$, $S = 5D/8$, $H = 4D$, $B = D/4$, and for the conditions stated in Fig. 3.13, Lapple (1951) gave a graphical solution of d_{50} [Eq. (3.19)] as a function of D and ρ_p. Note that constant values of N_t, v_i, and μ were assumed in these calculations. Hence, as D is varied, $Q = abv_i \propto D^2 v_i$ changes as D^2. Thus, in this case (cf. Eq. 3.12b), this scale-up relationship follows from Eq. (3.12a):

$$(d_1/d_2)_{50} = \sqrt{D_1/D_2} \tag{3.12c}$$

Lapple gave corrections of d_{50} for different values of v_i, μ, b/D, and N_t in Figs. 3.14 and 3.15.

For a cyclone having the proportions considered by Lapple, we have $V = 2.0125D^3$, $Q = 0.125D^2 v_i$, and thus, by Eq. (3.20),

$$N_t = \frac{(2.0125D^3/0.125D^2 v_i)\, v_i}{\pi D} = 5.12$$

It is apparent, then, that Eq. (3.20) predicts N_t to be independent of v_i.

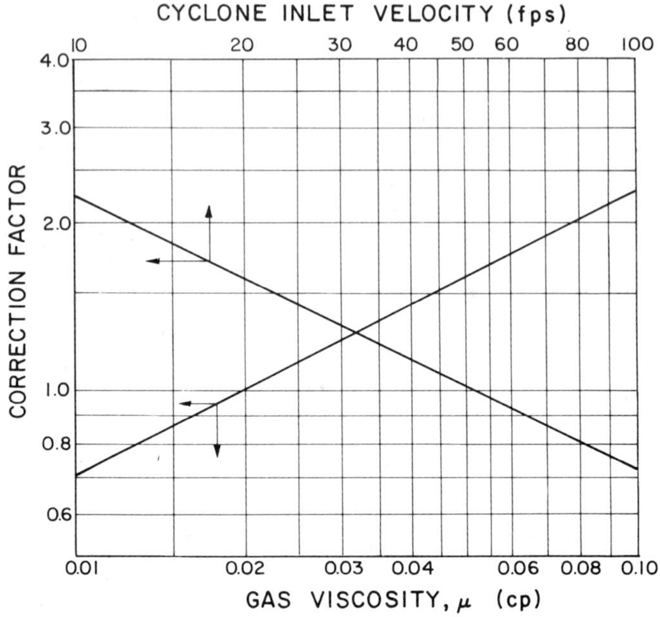

Figure 3.14 Viscosity and velocity correction factors for cut size particle of conventional cyclones (Lapple, 1951).

For cyclones of the proportions considered by Lapple, a generalized grade efficiency curve has been calculated (see Fig. 3.16). Grove (1971) has shown that the overall efficiency, η_T, of this type of cyclone may be estimated by the simple equation

$$\eta_T = \frac{\bar{d}}{\bar{d} + d_{50}} = \frac{1}{1 + (d_{50}/\bar{d})} \tag{3.22}$$

where d_{50} is calculated by Eq. (3.19), and $\bar{d}$ is the mass median particle diameter.

The effect of operating parameters on the efficiency of a given cyclone is

$$\frac{\text{Uncollected dust in cyclone 1}}{\text{Uncollected dust in cyclone 2}} = \frac{100 - \eta_1}{100 - \eta_2} = \sqrt{\frac{Q_2}{Q_1} \times \frac{\rho_{p2}}{\rho_{p1}} \times \frac{\mu_1}{\mu_2}} \times \left(\frac{L_{i,2}}{L_{i,1}}\right)^{0.182} \tag{3.23}$$

where L_i is the dust loading. Equation (3.23) predicts the same dependence

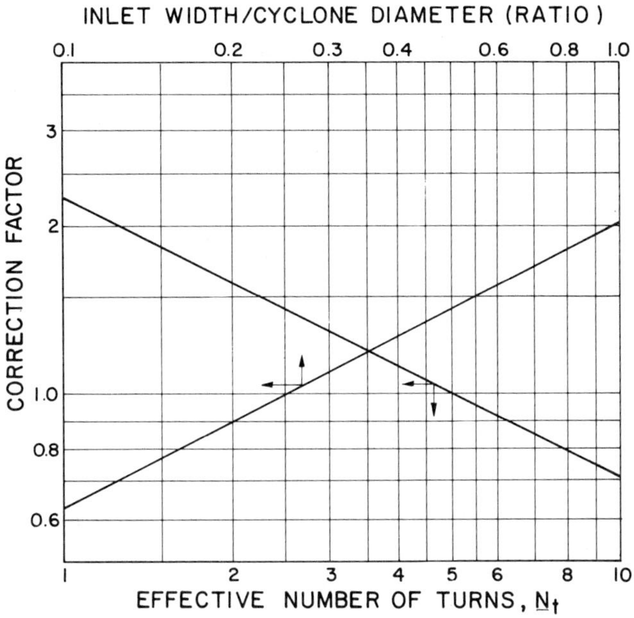

Figure 3.15 Inlet width/cyclone diameter and effective number of turns correction factors for cut size particle of conventional cyclones (Lapple, 1951).

Figure 3.16 Cyclone efficiency as a function of particle size ratio (Lapple, 1951).

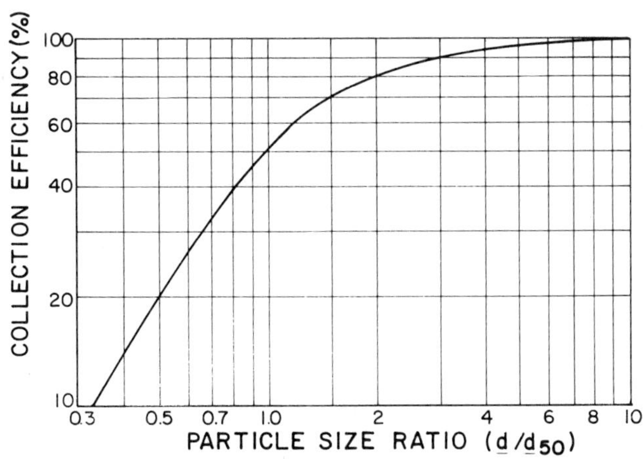

of the uncollected dust on Q, ρ_p, and μ as that predicted for d_{50} by Eq. (3.12b) for a given cyclone.

If the diameter of the cyclone is a variable, then this parameter has the greatest effect on the collection efficiency. Generally, many little cyclones in parallel produce better results than one big cyclone. The number of g's (n) in cyclones varies from 5 to 2,500; n is also called the separation factor. Inlet velocities range from 20 to 100 ft/s.

Method of Leith and Licht

Leith and Licht (1972) derived this equation for predicting cyclone efficiency:

$$\eta = 1 - \exp\left[-2(C\Psi)^{1/(2\gamma+2)}\right] \tag{3.24}$$

where C is a geometrical parameter:

$$C = \frac{\pi D^2}{ab}\left\{2\left[1 - \left(\frac{D_e}{D}\right)^2\right]\left(\frac{S}{D} - \frac{a}{2D}\right) + \frac{1}{3}\left(\frac{S + l - h}{D}\right)\left[1 + \frac{d}{D} + \left(\frac{d}{D}\right)^2\right] + \frac{h}{D} - \left(\frac{D_e}{D}\right)^2\frac{l}{D} - \frac{S}{D}\right\} \tag{3.25}$$

with

$$\frac{l}{D} = 2.3\frac{D_e}{D}\left(\frac{D^2}{ab}\right)^{1/3} \tag{3.26}$$

and

$$\frac{d}{D} = \frac{D - (D - B)[(S + l - h)/(H - h)]}{D}$$

Here, Ψ is a modified inertia parameter [see Eq. (4.10)]:

$$\Psi = \frac{\rho_p d_p^2 v_i}{18\mu D}(\gamma + 1) \tag{3.27}$$

and γ is the vortex exponent, given by Alexander (1949):

$$\gamma = 1 - \left[1 - \frac{(39.4D)^{0.14}}{2.5}\right]\left(\frac{T}{283}\right)^{0.3} \tag{3.28}$$

Here, D is the cyclone diameter in meters and T is the absolute temperature in Kelvin.

Leith and Licht (1972) have compared the methods of Barth (1956), Lapple (1951), and Sproull (1970) and their own method with the experimental grade efficiency curves of Stairmand (1951) and Peterson and

Whitby (1965). They have found their method to result in the best agreement with the experimental data.

Cyclone Pressure Drop

Generally speaking, if everything else is equal, more efficient cyclones result in higher pressure drops and, therefore, higher power consumption. It is just as important to be able to estimate the expected pressure drop as it is to measure the efficiency of the cyclone. There are various empirical relations that may be used to estimate the pressure drop in a cyclone, none of which is perfect.

Stairmand's Method

Stairmand (1949) gave this expression for the total pressure loss in a cyclone:

$$\Delta p = \frac{\rho}{2}\left(v_i^2\left\{1 + 2\phi^2\left[\frac{2(D-b)}{D_e} - 1\right]\right\} + 2v_e^2\right) \tag{3.29}$$

In this equation, v_e is the velocity in the exit pipe, that is,

$$v_e = 4Q/\pi D_e^2 \tag{3.30}$$

The pressure loss as expressed by Eq. (3.29) is composed of three parts: the entry pressure loss, the pressure loss inside the cyclone, and the exit pressure loss. The term containing the friction loss factor ϕ takes care of the friction losses inside the cyclone. As pointed out, the value of ϕ can either be calculated by Eq. (3.13) or read from the graph in Fig. 3.8.

Barth's Method

Another method of calculating the pressure loss in a cyclone that was given by Barth (1956), who expressed the pressure loss in terms of the entrance velocity and a dimensionless "pressure loss factor" ζ, is

$$\Delta p = \zeta(\rho v_i^2/2) \tag{3.31}$$

The factor ξ has been expressed by Barth as a function of the so-called "loss number" ε defined by

$$\varepsilon = \frac{\zeta}{(ab/\pi D_e^2/4)^2 (v_{\mathrm{Tmax}}/v_e)^2} \tag{3.32}$$

where ab is the cross-sectional area of the entrance pipe.

Both ε and ζ have been determined experimentally for a number of different cyclone types. These data, along with various other pertinent geometrical and physical properties of these cyclones, are listed in Table

3.4. The cyclone types A, D, and E are the same as those listed in Table 3.3. Nevertheless, the characteristic properties sometimes differ considerably for the same type of cyclone covered in the two tables, because the same cyclone type does not mean that all the relative proportions are the same.

Once the pressure loss factors, ζ, has been determined, the pressure loss in the cyclone can be calculated directly by using Eq. (3.31). In the absence of measured pressure loss factors, ζ in Eq. (3.31) is obtained via Eq. (3.32), as given next.

The loss number has been split into two components:

$$\varepsilon = \varepsilon_i + \varepsilon_e \tag{3.33}$$

and these equations have been developed (Barth, 1956) to find ε_i and ε_e:

$$\varepsilon_i = \frac{D_e}{D}\left\{\frac{1}{\left[1 - \dfrac{2v_{Tmax}(H - S)\mu'}{v_e D_e}\right]^2} - 1\right\} \tag{3.34}$$

Figure 3.17 (a) Alternative cyclone entry designs for loss factor calculations (Barth, 1956). (b) Values of loss factor α for pattern (ii) when $\alpha < 1$ (ter Linden, 1953). (c) Weidner's experimental curves for pressure loss factor (Weidner, 1954).

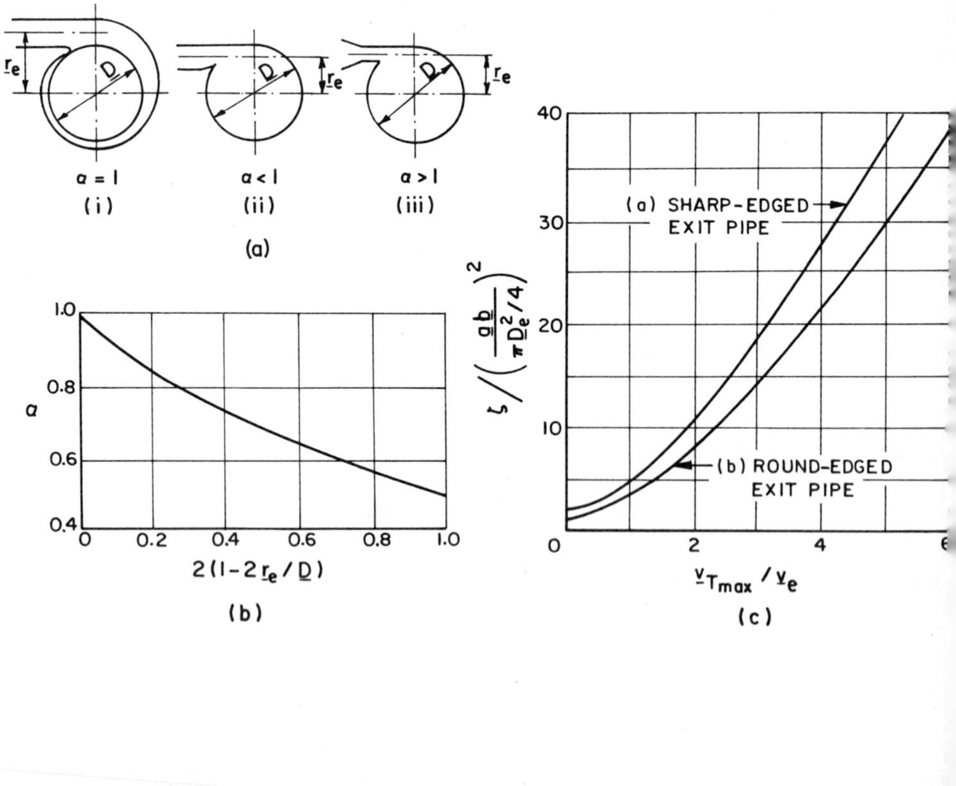

where μ' is the coefficient of friction (gas and wall) frequently assumed to be 0.02, and

$$\varepsilon_e = \frac{K}{(v_{\mathrm{Tmax}}/v_{\mathrm{e}})^{2/3}} + 1 \tag{3.35}$$

The parameter K in Eq. (3.35) has the values of 4.40 and 3.41 for sharp-edged and round-edged exit pipes, respectively.

Finally,

$$\frac{v_{\mathrm{Tmax}}}{v_{\mathrm{e}}} = \frac{\pi(D_{\mathrm{e}}/2)(D-b)}{2ab\alpha + \pi(H-S)(D-b)\mu'} \tag{3.36}$$

The entrance design loss factor α depends on the type of entry used. Three types are shown in Fig. 3.17(a). In Fig. 3.17(b), α is shown as a function of an entrance parameter $r_{\mathrm{e}} = (D-b)/2$, for $\alpha < 1$. Weidner

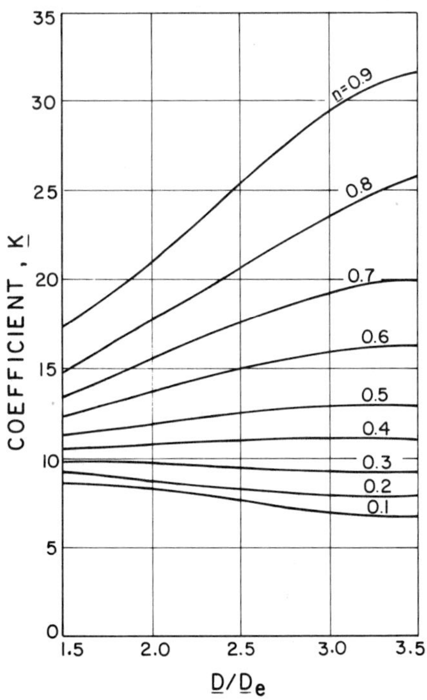

Figure 3.18 Value of K in Eq. (3.37) (Shepherd and Lapple, 1939, 1940).

(1954) measured $\zeta/[ab/(\pi D_e^2/4)]^2$ for a series of values of the ratio v_{Tmax}/v_e, and his data are plotted in Fig. 3.17(c). Curve (a) applies to sharp-edged exit pipes, curve (b) to round-edged ones. Once the pressure loss factor has been determined, the pressure drop in the cyclone can be immediately obtained by Eq. (3.27).

Method of Shepherd and Lapple

Experimental pressure drop results were correlated by Shepherd and Lapple (1939, 1940) with this equation:

$$\Delta p = K \frac{ab}{D_e^2} \frac{v_i^2 \rho}{2} \tag{3.37}$$

K is plotted versus D/D_e and n in Fig. 3.18, and n is plotted versus $D/2$ in Fig. 3.19; Δp may vary from 1 to 20 inlet velocity heads, depending on the geometrical proportions of the cyclone.

Figure 3.19 Values of n in Fig. 3.18 (Shepherd and Lapple, 1939, 1940).

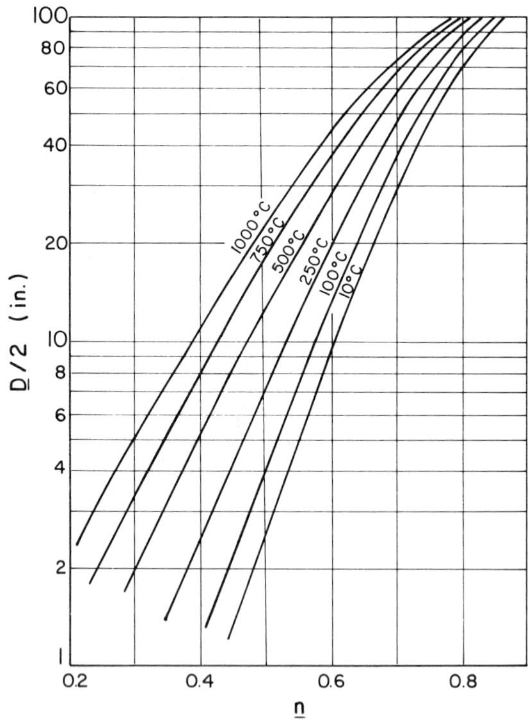

The effect of operating variables on the pressure loss of a given cyclone is

$$\frac{\Delta p_1}{\Delta p_2} = \frac{(Q_1^2 \rho_1/T_1)}{(Q_2^2 \rho_2/T_2)} \times \frac{0.013L_{i,2}^{0.5} + 1}{0.013L_{i,1}^{0.5} + 1} \tag{3.38}$$

where L_i is the inlet dust loading in grains/ft^3.

Leith and Mehta (1973) have compared the pressure drop equations of Straimand (1949), Barth (1956), Shepherd and Lapple (1939, 1940), First (1950), and Alexander (1949), using the measured values obtained by Shepherd and Lapple (1940), Stern *et al.* (1955), and Stairmand (1949). From their results, it appears that the Barth, Stairmand, and Shepherd and Lapple methods are best, on the average, even though all methods fall far short of perfection.

Optimal Cyclone Design

More recently several attempts have been made to optimize cyclone performance. The result of two different optimization calculations are now presented.

Optimal Cyclone Design of Leith and Mehta

Lengthy and careful considerations of cyclone performance that cannot be reproduced here led Leith and Mehta (1973) to express cyclone performance by the nondimensional parameter CD^2/ab:

$$\frac{CD^2}{ab} = \pi P^2\left(\frac{D_e}{D}\right)^4 \left\{ 2\left[1 - \left(\frac{D_e}{D}\right)^2\right]\left[\frac{H}{D} - 2.3P^{1/3}\left(\frac{D_e}{D}\right)^{5/3} - \frac{1}{P(D_e/D)^2(1 - (D_e/D))}\right] + \frac{1}{3}\left(\frac{H-h}{D}\right)\left[1 + \frac{B}{D} + \left(\frac{B}{D}\right)^2\right] + \left[1 - \left(\frac{D_e}{D}\right)^2\right]\left[2.3P^{1/3}\left(\frac{D_e}{D}\right)^{5/3}\right] + \frac{h}{D} - \frac{H}{D} \right\} \tag{3.39}$$

where P is a pressure drop coefficient:

$$P \equiv \Delta p D^4/8\rho Q^2 \tag{3.40}$$

Collection efficiency increases with increasing C [see Eq. (3.24)]. CD^2/ab and hence collection efficiency can be made as large as desired by increasing H/D and h/D. However, there are pracatical limitations on these quantities and, therefore, these were set at the values:

$$H/D = 5 \tag{3.41}$$

$$h/D = 3 \tag{3.42}$$

Also,

$$B/D = 0.375 \tag{3.43}$$

was set as a typical value. Substituting Eqs. (3.41)–(3.43) into Eq. (3.39) yields an expression for CD^2/ab that depends on D_e/D and P only, for the ratios S/D, b/D, and a/D have been expressed as functions of D_e/D and P as

$$\frac{S}{D} = \frac{H}{D} - 2.3\left[P\left(\frac{D_e}{D}\right)^5\right]^{1/3} \tag{3.44}$$

$$\frac{b}{D} = \frac{1}{2}\left(1 - \frac{D_e}{D}\right) \tag{3.45}$$

$$\frac{a}{D} = \frac{2}{P(D_e/D)^2[1 - (D_e/D)]} \tag{3.46}$$

Values of CD^2/ab were maximized on a digital computer for a range of values of P, by a trial-and-error procedure for D_e/D. With P fixed, the value of D_e/D leading to the greatest value of CD^2/ab was found. The other ratios, S/D, b/D, and a/D were calculated as functions of D_e/D from Eqs. (3.44)–(3.46), respectively. The consequent value of "optimized" C was then determined from Eq. (3.24).

The results of this optimization are shown in Figs. 3.20 and 3.21. Figure 3.20 is a plot of optimized C against P. Figure 3.21 shows optimized C against its corresponding dimension ratios D_e/D, S/D, b/D, and a/D. For any selected values of pressure drop Δp, diameter D, gas density ρ, and volume throughput Q, a value of P can be calculated from Eq. (3.40). Figure 3.20 will then provide the maximum value of C for this value of P. The dimension ratios for the cyclone can be found from Fig. 3.21. The efficiency of the cyclone for any particular dust and gas can be calculated from Eq. (3.24). The gas inlet velocity, v_i, is found as a function of the throughput, Q, and inlet height and width, through the relation $v_i = Q/(ab)$.

Optimal Cyclone Design by Koch and Licht

The cyclone collection efficiency η for particles of specific size and density was expressed by Koch and Licht (1977) as

$$\eta = 1 - \exp\left\{-2.0\left[\frac{G\tau Q(1+\gamma)}{D^3}\right]^{0.5/1+\gamma}\right\} \tag{3.47}$$

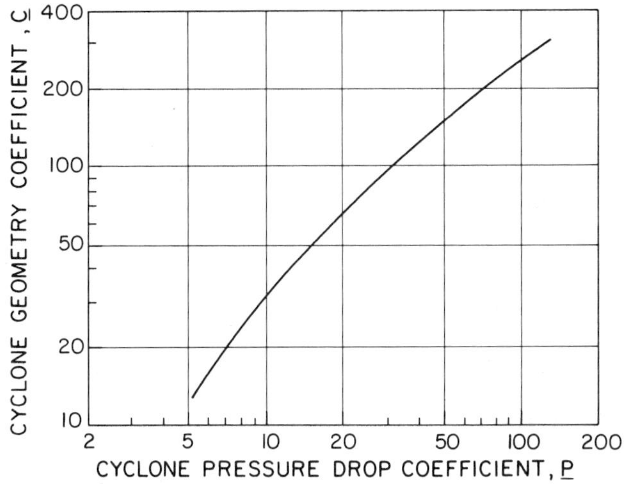

Figure 3.20 Relationship between cyclone pressure drop coefficient and geometry coefficient. [Reprinted with permission from *Atmospheric Environment*, **7**, D. Leith and D. Mehta, Cyclone performance and design, Copyright (1973), Pergamon Journals, Ltd.]

Figure 3.21 Cyclone dimension ratio versus cyclone geometry coefficient. [Reprinted with permission from *Atomospheric Environment*, **7**, D. Leith and D. Mehta, Cyclone performance and design, Copyright (1973), Pergamon Journals, Ltd.]

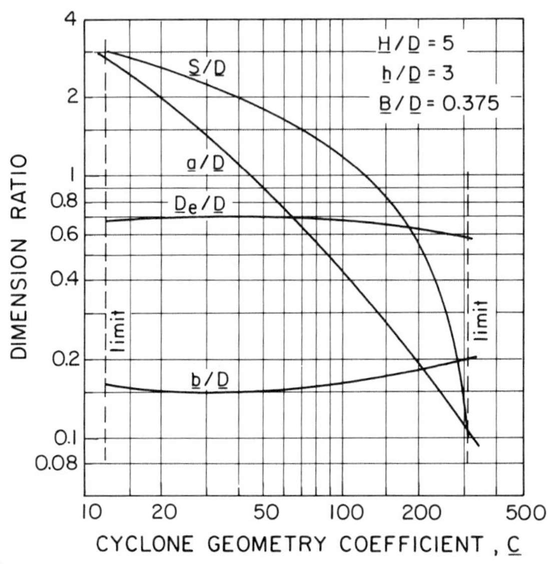

Table 3.5 Dimensionless Design Ratios for Tangential Entry Cyclones

		High efficiency		Conventional	
Symbol	Nomenclature	Stairmand (1951)	Swift (1969)	Lapple (1967)	Swift (1969)
D	Body diameter	1.0	1.0	1.0	1.0
a	Inlet length	0.5	0.44	0.5	0.5
b	Inlet width	0.2	0.21	0.25	0.25
S	Outlet length	0.5	0.5	0.625	0.6
D_e	Outlet diameter	0.5	0.4	0.5	0.5
h	Cylinder length	1.5	1.4	2.0	1.75
$H-h$	Cone length	2.5	2.5	2.0	2.0
H	Overall height	4.0	3.9	4.0	3.75
B	Bottom diameter	0.375	0.4	0.25	0.4
G	Configuration factor	551.3	699.2	402.9	381.8

where G is a cyclone configuration factor given in Table 3.5, γ is the vortex exponent given by Eq. (3.28), and τ is the relaxation time. This method of cyclone design yields the diameter, D, of a cyclone of fixed dimension ratios for maximum efficiency, given the throughput.

The ratio of inlet velocity to saltation velocity v_i/v_d has been shown to be an important factor in optimizing high-efficiency cyclone design. In the case of cyclones, the saltation velocity is the minimum gas velocity required to prevent the dropping or settling out of particles. Particle removal efficiency is a maximum in cyclones when the ratio $(v_i/v_d)_{\text{base}}$ equals 1.25, according to Kalen and Zenz (1974). Zenz (1964) further shows that reentrainment occurs when $v_i/v_d \geq 1.35$. Base conditions are 38°C and $\rho_p = 2.58\ \text{g/cm}^3$. To obtain the actual v_i/v_d, the base ratio is corrected for temperature and particle density variations by

$$\frac{v_i}{v_d} = \frac{(v_i/v_d)_{\text{base}}}{f_t f_\rho} \tag{3.48}$$

where

$$f_t = \frac{6.88}{10.52 - \ln t} \tag{3.49}$$

and

$$f_\rho = \frac{2.25}{3.11 - \ln \rho_p} \tag{3.50}$$

The temperature, t, is in °C and particle density, ρ_p, is in g/cm^3.

Koch and Licht (1977) have shown that

$$v_d = 0.228 v_a \left[\frac{(b/D)^{0.4}}{(1 - b/D)^{1/3}} \right] D^{0.067} v_i^{2/3} \tag{3.51}$$

where, according to Kalen and Zenz (1974) v_a, in cm/s, is given as

$$v_a = 1.10 \left[\frac{g\mu(\rho_p - \rho)}{\rho^2} \right]^{1/3} \tag{3.52}$$

According to this method, optimal cyclone design proceeds as follows. The gas throughput, Q; the dust particle size distribution; the temperature, t; the particle density, ρ_p; and the inlet dust loading, L_i are specified.

From the criterion $(v_i/v_d)_{\text{base}} = 1.25$, using the given values of t and ρ_p, v_i/v_d is obtained by Eqs. (3.48)–(3.50). Next, a certain cyclone configuration is selected from Table 3.5. The optimal cyclone diameter D depends, in principle, on Q, for it has been found by experience that small cyclones tend to perform more economically than large ones. Hence, it may be of advantage to split the throughput into ν parallel streams and feed each stream into a separate smaller cyclone than feed the entire throughput into a single large cyclone. Hence, the number of cyclones in parallel, ν, is introduced into the design problem as an additional variable by

$$\frac{Q\ (\text{m}^3/\text{s})}{\nu v_i\ (\text{cm/s})} 100\ (\text{cm/m}) = \frac{a}{D}\frac{b}{D} D^2\ (\text{m}^2) \tag{3.53}$$

whence

$$D\ (\text{m}) = 10 \left[\frac{Q}{(a/D)(b/D)} \right]^{1/2} \frac{1}{(\nu v_i)^{1/2}} \tag{3.54}$$

Equation (3.54) contains the two unknowns D and v_i, with ν as a parameter.

Equations (3.51) and (3.52) are combined to give v_d as a function of D and v_i; and then the expression of D, given by Eq. (3.54), is used to eliminate D. The resulting expression of v_d contains only the unknown v_i (in addition to the parameter ν). Then the ratio v_i/v_d is formed and is equated to the value obtained previously by Eqs. (3.48)–(3.50), yielding an equation containing only the unknown v_i (in addition to ν). Once v_i has been obtained, D is calculated by Eq. (3.54), and then the efficiency is obtained by Eq. (3.47). Here, v_i turns out to be a weak decreasing function of ν $[(v_i)_\nu = (v_i)_1/\nu^{0.09}]$, which means that the smaller the gas throughput per cyclone, the smaller is the optimal value of v_i. The optimal cyclone size D decreases with ν as $D = D_1/\nu^{0.455}$, resulting in a higher optimal efficiency if the gas throughput per cyclone is decreased, that is, if the number of cyclones used to treat the given gas flow is increased. As the pressure loss, Δp, varies proportionately with v_i^2, the optimal method predicts only a mild reduction in Δp for decreasing the gas throughput per cyclone, that is, increasing the number of cyclones to heat the given gas flow. If a certain fixed efficiency is stipulated, along with the rest of the conditions given in the design problem, other methods of cyclone design result in these relations (see Example 3.3)

$$D_\nu = D_1/\nu^{1/3} \tag{3.55}$$

and

$$\Delta p_\nu = \Delta p_1/\nu^{2/3} \tag{3.56}$$

where D_1 and Δp_1 are the cyclone diameter and pressure drop, respectively, if the entire throughput, Q, is treated by one cyclone; D_ν and Δp_ν are the corresponding values when Q is divided into ν parallel streams and each stream is fed into a separate cyclone to meet the same efficiency requirement as obtained when using only one cyclone.

EXAMPLE 3.3

Design a cyclone of type E in Table 3.3, handling a gas flow $Q = 200\ \text{m}^3\ \text{s}^{-1}$. The gas is air at 1 atm pressure and 90°C. A cut diameter of 8 μm is desired. The dust loading is 20 grains/ft^3. These additional data are known:

$$(a \times b)(\pi D_e^2/4) = 0.44; \qquad D/D_e = 4; \qquad 2b/D = 0.17;$$

$$2H/D_e = 7.5; \qquad S/D = 0.75; \qquad v_{Tmax}/v_i = 4.3;$$

$$\rho_p = 3 \times 10^3\ \text{kg m}^{-3}; \qquad \mu = 2.64 \times 10^{-5}\ \text{Pa s};$$

$$h = S; \qquad \Gamma = 0.007; \qquad \alpha = 0.93; \qquad \zeta = 37$$

1. Calculate the cyclone diameter D by both Barth's and Stairmand's methods.
2. Calculate the pressure loss Δp in the cyclone. Use both Stairmand's and Barth's methods. Analyze the various approaches for possible inconsistencies.
3. Let us suppose that the gas stream is divided into ν equal parallel streams, such that each stream will contribute a flow $Q_\nu = Q/\nu = 200/\nu$ (m^3 s^{-1}). Each stream is to be fed into a separate cyclone of type E. Derive and state the relationship between D_ν and D, where D_ν

is the diameter of each cyclone handling $200/\nu$ ($m^3\ s^{-1}$) of gas. Also derive and state the relationship between Δp_ν and Δp, where Δp_ν is the pressure loss in each cyclone of diameter D_ν, handling a flow Q_ν.

4. Let us assume that the power consumption is limited by the cost of power to 1 kW h/1000 m^3 gas. Calculate Δp_ν, ν, and D_ν, meeting this condition.
5. If 1 kW h costs 3 cents, how much does the operation of the ν cyclones of diameter D_ν, calculated under 4., cost the company every year? How much would the cost be if they used one big cyclone of diameter D, instead of the ν small ones of diameter D_ν?

SOLUTION

1. a. Barth's method:

$$\frac{u_t}{u_t^*} = \frac{d_{50}^2(\rho_p - \rho)\pi(H - S)v_{\mathrm{Tmax}}^2}{9\mu Q} = 1.25$$

(from Fig. 3.11, curve c, at 50% efficiency)

$$H = 7.5D_e/2 = 7.5D/(4 \times 2) = 7.5D/8$$

$$v_{\mathrm{Tmax}}^2 = 4.3^2 v_i^2 = 4.3^2 Q^2/a^2b^2$$

$$ab = 0.44\pi D_e^2/4 = 0.44\pi D^2/(4)^3$$

$$(ab)^2 = (0.44)^2\pi^2 D^4/(4)^6$$

$$v_{\mathrm{Tmax}}^2 = 4.3^2 Q^2 (4)^6/(0.44)^2\pi^2 D^4$$

$$\frac{u_t}{u_t^*} = \frac{8^2 \times 10^{-12} \times 3 \times 10^3\pi(7.5D/8 - 0.75D)4.3^2 \times 200^2(4)^6}{0.44^2\pi^2 D^4 \times 9 \times 2.64 \times 10^{-5} \times 200}$$

$$= 1.25$$

$$\frac{3773.36}{D^3} = 1.25$$

$$D = 14.45 \text{ m}$$

b. Stairmand's method:

$$d_{50} = \frac{3}{v_i\phi}\sqrt{\frac{\mu Q D_e}{2\pi\rho_p(H - S)D}}$$

where

$$\phi = \frac{-\sqrt{D_e/2(D - b)} + \sqrt{D_e/2(D - b)} + \sqrt{4\Gamma A/ab}}{2\Gamma A/ab}$$

and

$$A = \pi Dh + \frac{\pi(D+B)}{2}\sqrt{\left[\frac{(D-B)}{2}\right]^2 + (H-h)^2} + 2\pi D_e S + \frac{\pi(D^2 - D_e^2)}{4}$$

$$A \cong \pi(0.75)D^2 + \frac{\pi D}{2}\sqrt{\frac{D^2}{4} + \left(\frac{7.5D}{8} - 0.75D\right)^2} + \frac{0.75\pi D^2}{4} + \frac{\pi}{4}\left(D^2 - \frac{D^2}{16}\right)$$

$$A = \pi D^2\left(0.75 + \frac{1}{2}(0.215)^{1/2} + 0.375 + 0.2344\right) = 5D^2$$

$$ab = \frac{0.44\pi D^2}{(4)^3}$$

$$\frac{\Gamma A}{ab} = \frac{(0.007)(5D^2)}{0.44\pi D^2}(4)^3 = 1.62$$

$$\frac{D_e}{2(D-b)} = \frac{D}{(4)(2)(D - 0.17D/2)} = 0.1366$$

$$\therefore \phi = \frac{-\sqrt{0.1366} + \sqrt{0.1366 + (4)(1.62)}}{(2)(1.62)} = 0.68$$

$$\therefore \frac{d_{50}^2\phi^2 Q^2(4)^6}{(9)(0.44)^2\pi^2 D^4} = \frac{\mu QD/4}{2\pi\rho_p(7.5D/8 - 0.75D)D}$$

$$\therefore \frac{(8)^2 10^{-12}(0.68)^2(200)^2(4)^6}{(9)(0.44)^2\pi D^4} = \frac{2.64 \times 10^{-5}(200)}{(2)(4)(3)(10^3)D(7.5/8 - 0.75)}$$

$$\therefore \frac{8.8577 \times 10^{-4}}{D^4} = \frac{1.1733 \times 10^{-6}}{D}$$

$$\therefore D = (7.55 \times 10^2)^{1/3} = 9.1 \text{ m}$$

2. a. Stairmand's method:

$$\Delta p = \frac{\rho}{2}\left\{v_i^2\left[1 + 2\phi^2\frac{2(D-b)}{D_e}\right] + 2v_e^2\right\}$$

$$v_e = \frac{4Q}{\pi D_e^2} = \frac{(4)^3 Q}{\pi D^2}, \qquad v_e^2 = \frac{(4)^6 Q^2}{\pi^2 D^4}$$

$$v_i^2 = Q^2(4)^6/(0.44)^2\pi^2 D^4$$

$$\phi^2 = 0.68^2$$

$$\frac{2(D-b)}{D_e} = \frac{1}{0.1366}$$

$$\therefore \Delta p = \frac{0.9737}{2}\frac{(200)^2(4)^6}{\pi^2(9.1)^4}\left[\frac{1}{0.44^2}\left(1 + 2 \times 0.68^2 \frac{1}{0.1366}\right) + 2\right]$$

$$= 5.2 \times 10^4 \text{ Pa}$$

b. Barth's method:

$$\Delta p = \zeta \frac{v_i^2 \rho}{2} = 37 \frac{Q^2 (4)^6 (0.9737)}{(2)(0.44)^2 \pi^2 D^4}$$

$$\Delta p = \frac{(37)(200)^2(4)^6(0.9737)}{(2)(0.44)^2 \pi^2 (14.45)^4} = 3.71 \times 10^4 \text{ Pa}$$

3. $$\frac{(d_{50})_1}{(d_{50})_\nu} = \sqrt{\left(\frac{D_1}{D_\nu}\right)^3 \left(\frac{Q_\nu}{Q_1}\right)}$$ (from Stairmand's or Barth's formula)

where now

$$(d_{50})_1 = (d_{50})_\nu$$

and

$$Q_\nu = \frac{Q_1}{\nu}$$

$$1 = \sqrt{\left(\frac{D_1}{D}\right)^3 \left(\frac{1}{\nu}\right)}$$

$$D_\nu = D_1 \nu^{-1/3}$$

Also,

$$\frac{\Delta p_1}{\Delta p_\nu} = \frac{Q_1^2 D_\nu^4}{D_1^4 Q_\nu^2}$$ (from Stairmand's or Barth's formula)

$$\frac{\Delta p_1}{\Delta p_\nu} = \frac{Q_1^2}{D_1^4}\frac{D_1^4 \nu^{-4/3}}{Q_1^2/\nu^2} = \nu^{2/3}$$

$$\Delta p_\nu = \Delta p_1 \nu^{-2/3}$$

4. $$\frac{1 \text{ kW h}}{1000 \text{ m}^3} = 3600 \frac{\text{W s}}{\text{m}^3} = 3600 \frac{\text{kg}}{\text{m s}^2} = 3.6 \times 10^3 \text{ Pa} = 3.6 \text{ kPa}$$

Thus, the maximum $\Delta p_\nu = 3.6$ kPa. (Note that this is about a 14.5-in. H_2O column, which is still quite high.) Hence, from Barth's method,

$$\nu = \left(\frac{\Delta p_1}{\Delta p_\nu}\right)^{3/2} = \left(\frac{3.71 \times 10^4}{3.6 \times 10^3}\right)^{3/2} = 33$$

and from Stairmand's method,

$$\nu = \left(\frac{5.2 \times 10^4}{3.6 \times 10^3} \right)^{3/2} = 55$$

The corresponding diameters, D_ν, are, from Barth,

$$D_\nu = \frac{D_1}{\nu^{1/3}} = \frac{14.45}{33^{1/3}} = 4.5 \text{ m}$$

and from Stairmand,

$$D_\nu = \frac{9.1}{55^{1/3}} = 2.4 \text{ m}$$

Note that the inlet velocities, v_{i}, are, from Barth,

$$v_{\mathrm{i}} = \frac{Q}{33ab} = \frac{(200)(4)^3}{(33)(0.44)\pi(4.5)^2} = 13.86 \text{ m s}^{-2} = 45.4 \text{ fps}$$

and from Stairmand,

$$v_{\mathrm{i}} = \frac{(200)(4)^3}{(55)(0.44)\pi(2.4)^2} = 29.2 \text{ m s}^{-1} = 96 \text{ fps}$$

5. The company's operating expenses are determined as follows. The yearly volume of gas treated is

$$(200)(3{,}600)(24)(365) = 6{,}307{,}200{,}000 \text{ m}^3$$

The yearly cost of power is

$$(0.03)\frac{\$}{\text{kW h}}\,\frac{1 \text{ kW h}}{1000 \text{ m}^3} \times 6{,}307{,}200{,}000 \text{ m}^3 = 189{,}216\,\frac{\$}{\text{yr}}$$

The yearly cost of power if only 1 cyclone were used is, from Barth's method,

$$(0.03)\frac{\$}{\text{kW h}}\,\frac{1 \text{ kW h}}{1000 \text{ m}^3}\,\frac{3.71 \times 10^4 \text{ Pa}}{3.6 \times 10^3 \text{ Pa}} \times 6{,}307{,}200{,}000 \text{ m}^3 = 1.95 \text{ million } \frac{\$}{\text{yr}}$$

and from Stairmand's method,

$$(1.95)(10^6)\frac{5.2 \times 10^4}{3.71 \times 10^4} = 2{,}733{,}100\,\frac{\$}{\text{yr}}$$

EXAMPLE 3.4

1. Design the high-efficiency, medium-throughput cyclone of Stairmand (see Figs. 3.9 and 3.10) by the method of Koch and Licht; that is, using Q, t, ρ_{p}, and $d_{50} = 1.4\ \mu\text{m}$, and the geometrical proportions of this cyclone, calculate D, v_{i}, and η (let $\nu = 1$) for maximum removal efficiency and compare the calculated values with the experimental values reported by Stairmand. Draw your conclusions.

2. Calculate Δp in cm w.c. for the calculated conditions by both the Koch and Licht and the Stairmand methods and compare the results.

SOLUTION

$$Q = 5450 \times 0.2^2 \frac{\text{m}^3}{\text{hr}} \frac{\text{hr}}{3600\ \text{s}} = 0.06 \frac{\text{m}^3}{\text{s}}$$

$$t = 20°\text{C}$$

$$\rho_p = 2\ \text{g/cm}^3$$

$$d_{50} = 1.4\ \mu\text{m}$$

1.
$$\frac{v_i}{v_d} = \frac{(v_i/v_d)_b}{f_t f_\rho} = \frac{1.25}{0.914 \times 0.93} = 1.47$$

$$f_t = \frac{6.88}{10.52 - \ln t°\text{C}} = \frac{6.88}{10.52 - \ln 20°\text{C}} = 0.914$$

$$f_\rho = \frac{2.25}{3.11 - \ln \rho_p} = \frac{2.25}{3.11 - \ln 2} = 0.93$$

$$\frac{Q\ (\text{m}^3/\text{s}) \times 100}{v_i\ (\text{cm/s})} = a \times b = 0.2 \times 0.5D^2\ (\text{m}^2)$$

$$\frac{6}{v_i} = 0.1D^2 \qquad \therefore D = \left(\frac{60}{v_i}\right)^{1/2}$$

$$v_a = 1.10\left(\frac{981 \times 1.8 \times 10^{-4} \times 2}{1.2^2 \times 10^{-6}}\right)^{1/3} = 68.9\ \text{cm/s}$$

$$\therefore v_d = 0.228 \times 68.9\left[\frac{(0.2)^4}{(0.8)^{1/3}}\right] D^{0.067} v_i^{2/3} = 8.9 D^{0.067} v_i^{2/3}$$

$$\therefore v_d = 8.9\left(\frac{60}{v_i}\right)^{1/2 \times 0.067} v_i^{2/3} = 10.2 v_i^{0.633}$$

$$\therefore \frac{v_i}{v_d} = \frac{v_i}{10.2 v_i^{0.663}} = \frac{v_i^{0.367}}{10.2} = 1.47$$

$$\therefore v_i = (1.47 \times 10.2)^{1/0.367} = 1600\ \text{cm/s} = 16\ \text{m/s}$$

$$\therefore D = \left(\frac{60}{1600}\right)^{1/2} = 0.194\ \text{m}$$

$$\tau = \frac{d^2 \rho_p}{18\mu} = \frac{1.4^2 \times 10^{-12} \times 2 \times 10^3}{18 \times 1.8 \times 10^{-5}} = 1.2 \times 10^{-5}\ \text{s}$$

$$n = 1 - (1 - 0.669 \times 0.48^{0.14}) = 0.60$$

$$\therefore \eta = 1 - \exp\left[-2.0\left(\frac{382 \times 1.2 \times 10^{-5} \times 0.06 \times 1.6}{0.1943}\right)^{0.5/1.6}\right] = 0.56$$

These results are in excellent agreement with the actual data, that is,

$v_i = 15\ \text{m/s}; \quad D = 0.2\ \text{m};\ \eta = 0.5$

2. a. Method of Koch and Licht:

$$\Delta p = 8.19 \times 10^{-3} \times 1.2 \times 10^{-3} \times 1600^2\left(\frac{0.1}{0.25}\right) = 10\ \text{cm w.c.}$$

b. Stairmand's method:

$$Q = v_i \times a \times b = v_e \times \frac{D_e^2\pi}{4}$$

$$\therefore\ v_e = \frac{v_i \times a \times b}{D_e^2\pi/4} = v_i\frac{0.1}{0.196} = 0.51v_i$$

Assuming that $\phi = 1$,

$$\therefore\ \Delta p = 16^2 \times \frac{1.2}{1}\left[\left(1 + \frac{2 \times 2 \times 0.8}{0.5} - 2\right) + 0.2 \times 0.51^2\right] = 837\ \text{Pa}$$

$$\approx 8\ \text{cm w.c.}$$

Again, the agreement is very good.

3.6 Analysis of Cyclone Design Factors

This section and Section 3.7 follow portions of the review by Caplan (1977) fairly closely. The efficiency and pressure drop of a cyclone both tend to increase with increasing length of the body, decreasing inlet width, or increasing ratio of body diameter to outlet diameter, while decreasing the body diameter. An increase in length of the cyclone body provides for a longer residence time of gas in the vortex and, therefore, for more revolutions or turns in the vortex. One of the most common defects leading to lower efficiency is reentrainment of dust into the vortex core from the region of the dust discharge zone. The longer the cyclone below the gas outlet, the greater the opportunity for reentrained dust to be precipitated out of the vortex core before it enters the gas outlet. The height of the main vortex zone should be at least 5.5 times the gas outlet diameter, preferably more, perhaps up to 12 times the outlet diameter.

Increasing the ratio of the body diameter to the gas outlet diameter results in an increased efficiency, up to a ratio of about 3. However, there is a corresponding increase in the pressure drop as this ratio increases, so that the optimum ratio would appear to be between 2 and 3.

Cyclones usually have a cone, but it is not necessary to have one. However, a cone does serve the practical function of delivering the dust to a central point for ease in disposal.

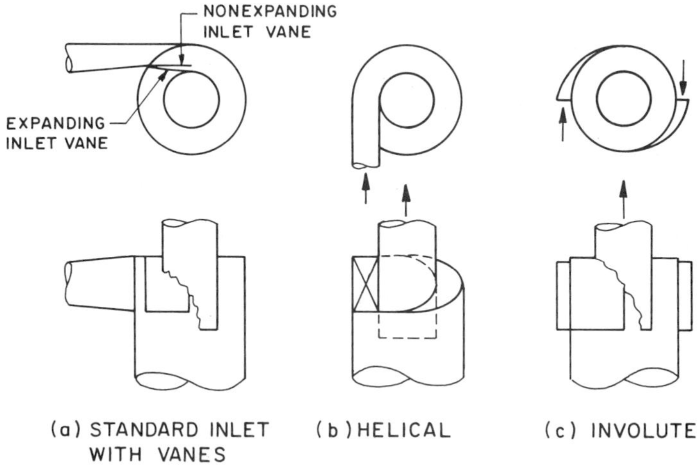

Figure 3.22 Types of tangential inlets.

The design of the cyclone inlet is of critical importance for the pressure drop. The different types of tangential inlets in use are shown in Fig. 3.22. The helical inlet design is provided to impart a downward velocity to the gas to avoid interference between the incoming gas and the gas already rotating in the annulus. Most commercial cyclones do not have a helical inlet.

As the inlet gas enters the annular space between the cyclone body wall and the wall of the gas outlet duct, it undergoes a squeeze between the body wall and the rotating air mass already in the annulus. The involute inlet design has been developed to minimize the interference between these gas streams.

The approach duct to a common cyclone is usually round, and if a proper inlet height-to-width ratio is to be obtained, the round duct must be transformed to a rectangular inlet. Such transformation should be gradual.

It is essential in any cyclone design to remove the separated dust from the cyclone cone or body as immediately, completely, and continuously as possible. The simple system is a hopper, or dust bin, closed at the bottom and opened at the top for the cyclone discharge. There is a vortex in the bin as well as in the core. A common method of minimizing upflow through the dust discharge pipe is to use some type of a valve to prevent such a flow. For most dust collection systems with a negative pressure at the bottom of the cyclone in the range of water gauge inches, ordinary rotary valves are sufficiently gas tight. If the negative pressure is higher, however, much better valving is required.

The presence of eddies in the annulus of the cyclone requires that the gas outlet have an extension into the body of the cyclone in order to minimize loss of dust through the gas outlet. The optimum length of the gas outlet extension is about one gas outlet diameter. In most instances, this extension terminates slightly below the gas inlet.

3.7 Effect of Operating Parameters on Cyclone Performance

Theoretically, the pressure drop varies with the square of the flow rate and, therefore, with the square of the inlet velocity. Experimentally, slight deviations from this behavior have been found; nevertheless, it is customary to assume that the pressure drop varies with the square of the flow rate.

Variation in the flow rate also has a marked effect on efficiency. Particles that are smaller than about 5 μm tend to follow the turbulent eddies so closely that they are not collected efficiently even under the influence of several hundred *g*'s usually present in cyclone collectors. Efficiency increases with increasing flow rate only up to some limiting velocity, above which reentrainment increases more rapidly than separation, thus causing a decrease in efficiency with a further increase in flow rate.

Plots of cyclone efficiency versus pressure drop are more useful for cyclone selection than are plots of efficiency versus flow. Typical curves of the various parameters discussed are shown in Figs. 3.23 and 3.24.

An interesting feature of cyclones is the fact that an increased load into a cyclone causes the pressure loss to be lower and the efficiency to be

Figure 3.23 Variation of efficiency with inlet velocity (Caplan, 1977).

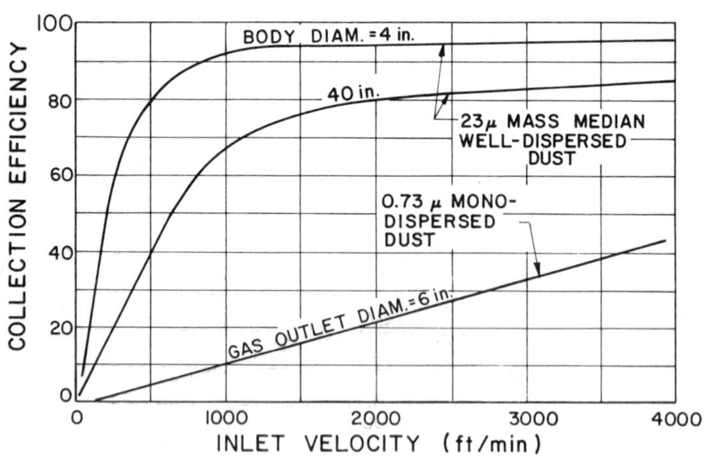

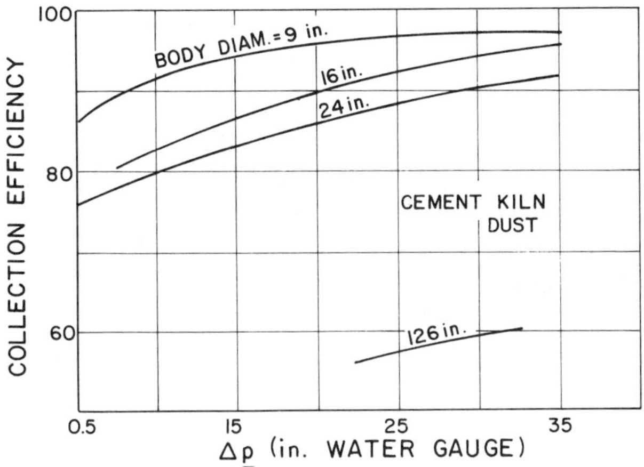

Figure 3.24 Variation of efficiency with pressure drop (Caplan, 1977).

increased. Apparently the pressure drop at a heavy dust load of 75–100 grains per cubic foot will be in the range of 75–85% of that with essentially clean air.

Dust collection efficiency increases with dust loading. At very high loadings, all cyclones have relatively high efficiencies, although the gas discharge still contains progressively higher dust loadings. Apparently the details of design become less important to efficiency as loading increases, but because erosion and plugging become more important factors, cyclones are designed to overcome those problems. Therefore, small diameter cyclones are seldom used for very heavy dust loads.

Where high efficiency cyclones are used, it is customary to operate a number of them in parallel to achieve a practical gas volume. If the number of cyclones in parallel is small, each cyclone should have its own inlet and its own dustbin. The inlet should provide relatively uniform distribution of gas and dust to each of the cyclones. The gas outlets may be direct to the atmosphere or may be manifolded together into a common duct.

When the number of cyclones in parallel is large, the only practical arrangement is to use a common inlet plenum chamber, a common dustbin, and a common outlet plenum chamber. Under such circumstances, new operating problems arise. To obtain efficiencies near those obtainable with a single cyclone (tube) of the same size, it is necessary to equalize the gas and dust load distribution to the cyclones to prevent backflow through individual cyclones or reentrainment from the dustbin. A typical arrangement of multiple, small-diameter, high-efficiency cyclones is shown in Fig. 3.25. In optimizing cyclone design and performance, the problem is the

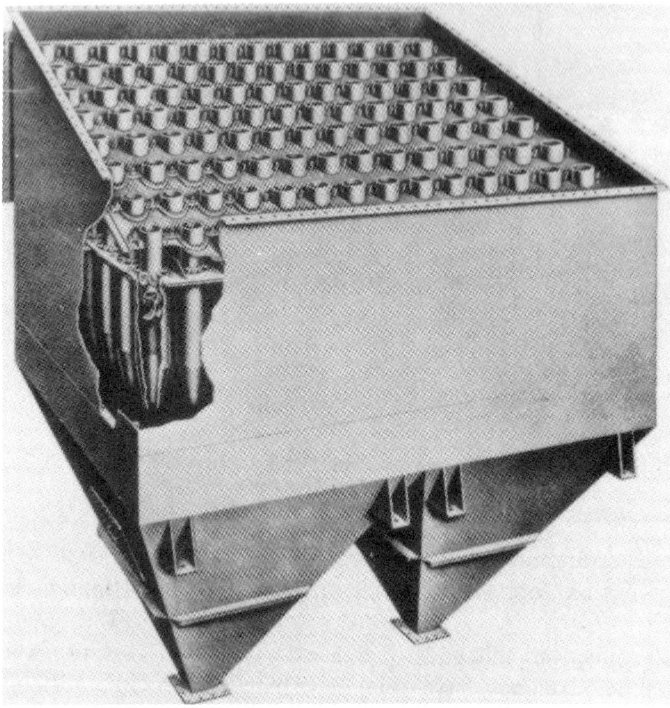

Figure 3.25 Typical bank of small-diameter, high-efficiency cyclones. (Courtesy Research-Cottrell, Inc.)

balance between pressure loss and efficiency. The objective is a cyclone design to achieve the required efficiency on a known dust distribution at minimum pressure loss.

PROBLEMS

1. In textbooks on air pollution control and equipment, problems on gravity settlers often create the impression that by allowing a sufficiently great residence time for the particles, very small, even submicron particles, can be collected with a very high efficiency in the settling chambers. The following is an example.

   The installation of a gravity settler has been proposed to help remove limestone particles (density equal to 2.7 g cm^{-3}) from air at ambient conditions. The inlet dust loading is 35 grains/ft^3, the volumetric flow rate is 980 ft^3/min. The inlet size distribution of the limestone is given in this table.

Size range (μm)	Weight (%)
0–5	2
5–10	4
10–20	12
20–60	40
40–80	28
80–120	10
>120	4

The chamber is 30 ft wide, 25 ft high, and 80 ft long. Calculate and plot the average collection efficiency in each size range and also calculate the overall collection efficiency.

Discuss critically the calculated efficiencies.

2. An inventor has modified the traditional shape of the settling chamber by removing the taper with respect to the horizontal direction from the transition piece connecting the chamber with the gas outlet duct. In the new design, the top of the chamber continues straight all the way to the outlet duct, and the end of the chamber consists of a vertical wall up to a height where the bottom of the outlet duct is situated. The idea is to make the dust particles that did not settle in the chamber to impinge on the vertical wall, deposit there, and then fall off the wall in chunks into the hopper.

   Repeat the calculation of problem 1 for this modified settling chamber if the height of the vertical wall is 20 ft and all the dust reaching the wall is removed from the gas stream.

   Discuss critically the invention and the results of your calculation.

3. a. You are requested to design a gravity settler to remove all the iron particulates from a dust-laden gas stream. This information is given: $d_p = 35$ μm (uniform, i.e., no distribution), gas = air at ambient conditions, $Q = 130$ ACFS, and $\rho_p = 7.62$ g cm^{-3} (assume $L = B$).

   b. After having completed the design of the gravity settler, you are now requested to repeat the calculations for all conditions the same as given in a, except the particle size is now bigger, that is, $d_p = 70$ μm. How did the dimensions of the gravity settler change when the particle size was doubled?

4. Example 3.1 in this chapter presents the calculation of the grade efficiency curve of a gravity settler of these dimensions:

   10 ft wide × 20 ft high × 45 ft long

   The inlet size distribution of the fly ash of 2.31 specific gravity is given in this table.

Size range (μm)	Mass (%)
0–10	2
10–20	2
20–30	6
30–40	19
40–60	20
60–80	22
80–100	18
100–150	11

   Do the calculation of the grade efficiency curve (part 1 of Example 3.1) for the volume flow of 100,000 ACFM, which is more appropriate for a settler of these dimensions. Check whether the assumption $u_m = u_t$ is also good in this case.

5. You are required to design a gravity settler to operate under these conditions:

   $$Q = 50{,}000 \text{ ACFM}$$

   $$\rho = 1.195 \times 10^{-3} \text{ g/cm}^3$$

   $$\mu = 1.84 \times 10^{-4} \text{ g/cm s}$$

   $$\rho_p = 6.0 \text{ g/cm}^3$$

   Cube-shaped particles

   $$d_v = 0.01 \text{ cm}$$

   $$L_i = 50 \text{ grain/ft}^3$$

   $$\eta = 1$$

   $$v \leq 6 \text{ m/s}$$

   You must decide on the settler volume and the actual dimensions of the settler.

6. Design a cyclone (single or multicyclone) handling a gas flow of 200 m^3 s^{-1} to obtain a cut diameter of 8 μm of a dust with a density equal to 3 g cm^{-3}. The gas is essentially air at 1 atm pressure and 90°C temperature. The inlet dust loading is 20 grains/ft^3. A maximum

pressure drop of 14.5-in. H_2O column is permitted. Base your design on the high-efficiency cyclone of Stairmand [Fig. 3.9(a)]. Use Eqs. (3.12b) and (3.29).

7. Repeat the design calculation of problem 6, but use the method of Leith and Licht, that is, Eqs. (3.24)–(3.28) and Eq. (3.37).

8. Repeat the design calculation of problem 6, but use the optimal cyclone design method of Leith and Mehta, that is, Figs. 3.20 and 3.21, resulting in using the dimension ratios $(H/D) = 5$ and $(h/D) = 3$, instead of $(H/D) = 4$ and $(h/D) = 1.5$.

9. Repeat the design calculation of problem 6, but use the optimal cyclone design method of Koch and Licht, that is, Eqs. (3.47)–(3.52).

10. Repeat the design calculation of problem 6, but base your design on the conventional cyclone of Lapple (see Table 3.5). Use Figs. 3.13–3.15 and Eq. (3.37).

11. Repeat the design calculation of problem 10, but use the method of Leith and Licht, that is, Eqs. (3.24)–(3.28).

12. Repeat the design calculation of problem 10, but use the optimal cyclone design method of Leith and Mehta, that is, Figs. 3.20 and 3.21, resulting in using the dimension ratios $(H/D) = 5$ and $(h/D) = 3$, instead of $(H/D) = 4$ and $(h/D) = 2$.

13. Repeat the design calculation of problem 10, but use the optimal cyclone design method of Koch and Licht, i.e., Eqs. (3.47)–(3.52).

14. Cascade impactor sampling on a cyclone treating a gas stream of viscosity equal to 0.030 cp, with a dust loading of 3 grains/ft^3, produced these results.

Size range (μm)	Weight (%) collected in inlet stream	Weight (%) collected in outlet stream
0–5	8.05	76.0
5–10	2.55	12.9
10–15	2.16	4.5
15–20	2.10	2.1
20–25	2.04	1.5
25–30	1.87	0.7
30–35	1.85	0.5
35–40	1.84	0.4
40–45	1.83	0.3
>45	75.71	1.1
	100.00	100.0

Stack sampling gave an overall removal efficiency of 90%. Plot the fractional efficiency curve and determine the cut diameter.

15. Consider a type D cyclone, the dimensions of which are given in Table 3.4; and

$$S = \frac{4}{3}a, \qquad D = 30 \text{ cm}, \qquad \zeta = 22$$

$$d = 7\ \mu\text{m}$$

$$\rho_p = 3\ \text{g/cm}^3$$

$$\mu = 1.82 \times 10^{-4}\ \text{g/cm s}$$

$$\rho = 1.2 \times 10^{-3}\ \text{g/cm}^3$$

$$L_i = 100\ \text{grains/ft}^3$$

a. Using Barth's Method and curve d in Fig. 3.11, calculate and tabulate these relationships:
   i. the efficiency versus the inlet velocity in cm s^{-1} and volume flow in $m^3\ s^{-1}$ and
   ii. the efficiency versus the pressure drop.

b. What is the cut diameter of the cyclone when the removal efficiency is
   i. 90%?
   ii. 10%?

c. Calculate the volume flow Q in a type D cyclone with a body diameter of 3 m that would collect the same dust under the same conditions with the same efficiency as the 30-cm diameter cyclone does for a volume flow of $1.374 \times 10^{-2}\ m^3\ s^{-1}$. How many small cyclones with a diameter of 30 cm would be required in parallel to handle this gas flow?

d. Calculate the pressure loss Δp in the 3-m-diam. cyclone of c. if the pressure loss in a 30-cm diameter cyclone is 0.605 cm water column.

e. Repeat the calculations in a through d by using the optimal cyclone design method of Koch and Licht, that is, Eqs. (3.47)–(3.52). (Use $G = 250$.)

f. Discuss the results.

16. A cyclone with a 3.0-m body diameter of the design considered by Lapple is to be used to remove particulates from a dust-laden gas at 20°C and 1 atm ($\mu = 1.82 \times 10^{-5}$ Pa s, $\rho = 1.2\ \text{kg/m}^3$). The dust

density is 3.0×10^3 kg/m^3. The particle size distribution of the dust is known.

a. As one of the steps in finding the best operating conditions of the cyclone, calculate the inlet velocity, in m s^{-1}, and the volume flow, in m^3 s^{-1}, to obtain a 50% removal efficiency of 7-μm diameter particles. Use the method recommended by Lapple.

b. For the inlet velocity just obtained, calculate the pressure loss in pascals (1 in. H_2O = 0.2491 kPa). The inlet loading is 100 grains/ft^3. Use the method recommended by Lapple. (1 kg/m^3 = 0.06243 lb/ft^3).

References

Alden, J. L. (1959). "Design of Industrial Exhaust Systems," 3rd Ed. The Industrial Press, New York.

Alexander, R. M. (1949). Fundamentals of cyclone design and operation. *Proc. Australas. Inst. Min. Metall.* **152–153**, 203–238.

Baliff, J., Greenburg, J., and Stern, A. C. (1948). *Am. Ind. Hyg. Assoc. Q.* **9**, 85.

Barth, W. (1956). Berechnung und Auslegung von Zyklonabscheidern. *Brennst. Wäerme Kraft* **8**, 1–9.

Caplan, K. J. (1977). Source control by centrifugal force and gravity. *In* "Air Pollution" (A. C. Stern, ed.), 3rd Ed., Vol. IV. Academic Press, New York.

First, M. W. (1950). Fundamental factors in the design of cyclone dust collectors. Doctoral thesis, Harvard University, Cambridge, Massachusetts.

Grove, D. J. (1971). Cyclones. *In* "Control of Particulate Emissions." U.S. Environmental Protection Agency, Washington, D.C.

Jennings, R. F. (1950). Blast-furnace gas cleaning-analysis of plant performance. *J. Iron Steel Inst.* **164**, 305–325.

Kalen, B., and Zenz, F. A. (1974). *AIQE Symp. Ser.* **70**(137), 388.

Koch, W. H., and Licht, W. (1977). New design approach boosts cyclone efficiency. *Chem. Eng. Int. Ed.* **84**, 80–88.

Lapple, C. E. (1951). Precipitators stop dust and fumes. *Chem. Eng. Int. Ed.* **58**, 144–151.

Lapple, C. E. (1967). *In* "Air Pollution Engineer Manual" (J. A. Danielson, ed.), p. 95. U.S. Dept. of Health, Education and Welfare, PHS No. 999-AP-40.

Leith, D., and Licht, W. (1972). "Collection Efficiency of Cyclone Type Particle Collectors; A New Theoretical Approach." *AIQE Symp. Ser.* **A12**-1971.

Leith, D., and Mehta, D. (1973). Cyclone performance and design. *Atmos. Environ.* **7**, 527–549.

Muschelknautz, E. (1967). Investigations on centrifugal separators. *Chem. Ing. Tech.* **39**, 306–310.

Muschelknautz, E., and Brunner, K. (1967). Experiments with cyclones. *Chem. Ing. Tech.* **39**, 531–538.

Peterson, C. M., and Whitby, K. T. (1965). Fractional efficiency characteristics of unit type collectors. *ASHRAE J.* **7**(5), 42–49.

Prockat, F. (1930). A contribution to the coal dust problem. *Glasers Ann.* **106**, 73–79, 93–97, 151–154; **107**, 38–45, 47–54.

Shepherd, C. B., and Lapple, C. E. (1939). Flow pattern and pressure drop in cyclone dust collectors. *Ind. Eng. Chem.* **31**, 972–984.

Shepherd, C. B., and Lapple, C. E. (1940). Flow pattern and pressure drop in cyclone dust collectors: Cyclone without inlet valve. *Ind. Eng. Chem.* **32**, 1246–1248.

Sproull, W. T. (1970). "Air Pollution and Its Control." Exposition Press, New York.

Stairmand, C. J. (1949). Pressure drop in cyclone separators. *Engineering* **168**, 409–412.

Stairmand, C. J. (1951). Design and performance of cyclone separators. *Trans. Inst. Chem. Eng.* **29**, 356–573.

Stairmand, C. J., and Kelsey, R. N. (1955). The role of the cyclone in reducing atmospheric pollution. *Chem. Ind. (London)*, 1324–1330.

Stern, A. C., Caplan, K. J., and Bush, P. D. (1955). Cyclone dust collectors. Amer. Petrol. Inst. Rep.

Strauss, W. (1975). "Industrial Gas Cleaning," 2nd Ed. Pergamon, Oxford.

Swift, P. (1969). *Steam Heat. Eng.* **38**, 453.

ter Linden, A. J. (1949). Investigations into cyclone dust collectors. *Proc. Inst. Mech. Eng.* **160**, 233–140.

ter Linden, A. J. (1953). *Tonindust. Ztg.* **22**, 49.

Weidner, G. (1954). *Ver. Deutsch. Ing. Tagungsh.* **3**, 16.

Zenz, F. A. (1964). Conveyability of materials of mixed particle size. *Ind. Eng. Chem. Fund.* **3**, 65.

4

Aerodynamic Capture of Particles

4.1 Introduction

Aerodynamic capture of particles plays a central role in the working of filters, spray towers, and scrubbers. In this type of equipment, the particles are captured by small collecting bodies, such as filter fibers or water droplets. In order for the capture of a particle to take place, there must first exist a relative velocity between the particle carried by the gas stream and the collecting body, and in addition, the particle must impinge on and adhere to the collecting body rather than bypass it. In this chapter, the mechanism of particle capture by collecting bodies is discussed and analyzed.

In order to simplify the complex mathematics of the capture of a particle by a collecting body, it has been customary to split the problem into two main parts. When tackling the first part, the particles are treated, from the point of view of contacting the collecting body, as mathematical points as shown in Fig. 4.1; whereas, in the second part, they are treated as if they did not have mass, only size, as shown in Fig. 4.2.

When the particles are treated as mass points traveling along streamlines, which they may leave when the streamlines approaching the collecting body start curving around the collecting body, the collection mechanism is called *inertial impaction*.

The particles are usually also considered as mathematical points when undergoing Brownian movement or diffusion.

However, the collection mechanism in which the finite size of a particle alone is considered to be responsible for the collection is called *interception*. The three different kinds of mechanisms, that is, inertial impaction, interception, and diffusion, are discussed in this chapter.

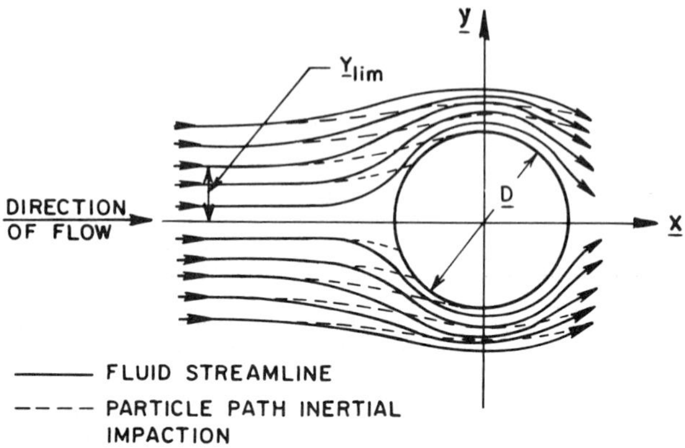

Figure 4.1 Flow pattern around cylinder, showing inertial impaction mechanism for spheres. [Reprinted with permission from Strauss (1975).]

4.2 Inertial Impaction

Streamline Patterns about a Cylinder and a Sphere

It is assumed that far from the collecting body the aerosol particles are traveling at the same velocity relative to the collecting body as the gas. As is well known from fluid mechanics, it is customary to represent the velocity of the fluid by streamlines. In steady flow, the direction of the fluid velocity

Figure 4.2 Flow pattern around cylinder, showing interception collection mechanism for spheres of differing sizes. [Reprinted with permission from Strauss (1975).]

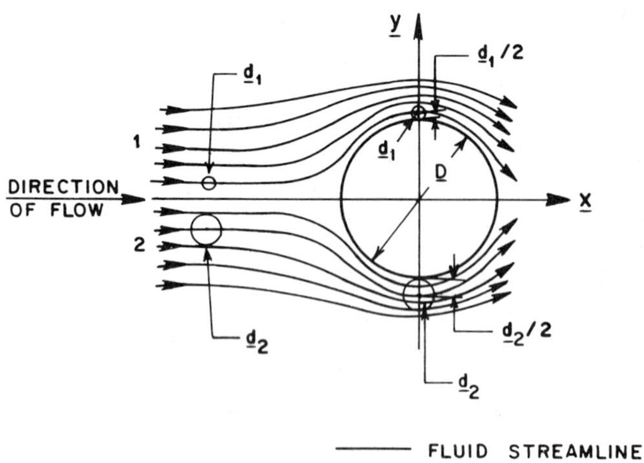

is always tangential to the streamlines at every point. Where the spacings between the streamlines are smaller, the fluid velocity is higher, and vice versa. In steady viscous (or creeping) flow, where Stokes' law is valid, the fluid velocity is at every point truly represented by the streamlines. In turbulent flow, however, only the mean fluid velocity is described by the streamlines, because the actual fluid velocity undergoes fluctuations about the mean at every point in space with a very high frequency. Nevertheless, we are often interested only in the average velocities, and so the streamline description of the flow pattern has been both popular and useful. Since in turbulent flow, sufficiently far away from any solid surface, inertial forces predominate over viscous forces, it has been customary to use, under these conditions, the so-called potential flow model in which the viscosity of the fluid is assumed to be equal to zero. A streamline pattern around a stationary cylinder, according to the potential flow model, is shown in both Figs. 4.1 and 4.2.

It has been assumed in these types of calculations that the aerosol particles travel along the streamlines, which is only true if diffusion can be neglected. As a result of diffusion, the particles keep changing streamlines, but the average number of particles on a streamline does not change as long as the particle concentration is uniform and the streamlines are straight.

For low collector Reynolds number (Re_c) [see Eq. (4.13)], where Stokes' law is valid, the viscous (creeping) flow model applies rigorously; whereas, for high values of the collector Reynolds number, the potential flow model is used. This gives the streamline in a good approximation, except in the boundary layer very near the surface of the collecting body. Naturally, there is a transition zone between low and high collector Reynolds numbers, where neither of the two models provides an adequate description of the flow.

The Particle Equations of Motion

As can be seen in Fig. 4.1, as the fluid flows around the collecting body, the streamlines become distorted. The aerosol particles may leave a curved streamline at a certain point upstream from the collecting body, because of inertia, and may subsequently impact against the surface of the collector.

The differential equation governing the motion of the particles is, in general, Eq. (2.25). In the situation considered here, $\mathbf{G}$ is equal to zero, because no external driving force acts on the particle. If the particle Reynolds number is small enough, then the resisting drag force is given by Stokes' law, which is written here in vectorial form:

$$\mathbf{F}_A = \frac{3\pi\mu d}{C}(\mathbf{w} - \mathbf{v}) \tag{4.1}$$

Here **w** and **v** are the particle and fluid velocities, respectively, taken *relative to the collector* (note that $\mathbf{u} = \mathbf{w} - \mathbf{v}$). Assuming spherical particles and steady flow, the combination of Eq. (4.1) with Eq. (2.25) yields

$$\mathbf{w} - \mathbf{v} = -\frac{C\rho_p d^2}{18\mu}\frac{d\mathbf{w}}{dt} \tag{4.2}$$

(note: in steady flow $d\mathbf{v}/dt = 0$.)

The mathematical problem consists of finding the trajectory of the particles, that is, the equations $x(t)$ and $y(t)$, from which one can calculate $y = f(x)$. In those special cases when the fluid velocity is either zero or constant, the solution of this differential equation is simple (see Example 4.1).

In the more general problem, however, the fluid velocity is given as a function of the position coordinates x and y by the model of the flow, that is, by the streamline pattern. Under these conditions, the solution is complicated and can be done only numerically, by using finite difference techniques.

The differential equation, Eq. (4.2), can be written in component form for the case of cylindrical collectors as

$$\left(\frac{d\tilde{x}}{d\tilde{t}} - \tilde{v}_x\right) = -\frac{2C\rho_p d^2 v_\infty}{18\mu D}\frac{d^2\tilde{x}}{d\tilde{t}^2} \tag{4.3}$$

and

$$\left(\frac{d\tilde{y}}{d\tilde{t}} - \tilde{v}_y\right) = -\frac{2C\rho_p d^2 v_\infty}{18\mu D}\frac{d^2\tilde{y}}{d\tilde{t}^2} \tag{4.4}$$

where these dimensionless variables have been introduced:

$$\tilde{x} = x/(D/2) \tag{4.5}$$

$$\tilde{y} = y/(D/2) \tag{4.6}$$

$$\tilde{v}_x = v_x/v_\infty \tag{4.7}$$

$$\tilde{v}_y = v_y/v_\infty \tag{4.8}$$

$$\tilde{t} = v_\infty t/(D/2) \tag{4.9}$$

Here, v_∞ is the velocity of the undisturbed fluid, that is, the velocity of the particle far from the collecting body, and D is the diameter of the collector. The parameter on the right-hand side of Eqs. (4.3) and (4.4) is the inertial impaction parameter Ψ, defined as

$$\Psi = \frac{C\rho_p d^2 v_\infty}{18\mu D} \tag{4.10}$$

Often twice the inertial impaction parameter, the Stokes' number is used:

$$\text{Stk} = \frac{C\rho_p d^2 v_\infty}{18\mu(D/2)} \tag{4.11}$$

that is, Stk = 2Ψ.

The "stopping distance" of a particle entering a still fluid with an initial velocity v_∞ in the Stokes' law region is (see Example 4.1)

$$x_s = v_\infty \tau = \frac{C\rho_p d^2 v_\infty}{18\mu} \tag{4.12}$$

It is apparent that division of this distance with D gives the inertial impaction parameter Ψ. Hence, the impaction parameter Ψ is the (dimensionless) stopping distance of the particle expressed in units of the collector diameter D. Similarly, the Stokes' number is the dimensionless stopping distance expressed in units of the collector radius $D/2$.

EXAMPLE 4.1

Consider the event of a spherical particle of velocity u_0 being catapulted horizontally into still air. What distance will the particle travel in the horizontal direction before the velocity u_0 decreases to zero, that is, what is the stopping distance of the particle?

SOLUTION

The force balance on the particle is

$$-\mathbf{F}_A = m\mathbf{a}$$

Assuming the validity of Stokes' law and considering the horizontal velocity component only,

$$-\frac{3\pi d u \mu}{C} = m\frac{du}{dt}$$

$$-\int_0^t \frac{3\pi d\mu\, dt}{mC} = \int_{u_0}^{u} \frac{du}{u}$$

$$\therefore\ u = u_0 e^{-t/\tau} \qquad \left(\tau = \frac{mC}{3\pi d\mu}\right)$$

But, $u = dx/dt$.

$$\int_0^{x_s} dx = u_0 \int_0^\infty e^{-t/\tau}\, dt$$

$$\therefore\ x_s = -u_0\tau\left(e^{-t/\tau}\right)_0^\infty = u_0\tau = \frac{u_0 mC}{3\pi d\mu}$$

As

$$m = \frac{d^3 \pi}{6} \rho_p$$

$$\therefore x_s = C \frac{u_0 d^2 \rho_p}{18 \mu}$$

Again, x_s is the stopping distance of the particle.

Collection Efficiency by Inertial Impaction Mechanism

As the streamline pattern varies with the value of the collector Reynolds number Re_c,

$$\mathrm{Re}_c = \frac{v_\infty D \rho}{\mu} \tag{4.13}$$

where D is the diameter of collector, the trajectory of a particle leaving a streamline at a certain point $(\tilde{x}, \tilde{y})$ will also depend on Re_c, in addition to the inertial impaction parameter Ψ.

It follows that the efficiency of collection of particles traveling along the streamlines far upstream from the collecting body is a function of two parameters: the inertial impaction parameter and the collector Reynolds number Re_c. The collection efficiency is calculated by finding the so-called "limiting streamline" in Fig. 4.1, which is the particular streamline from which the trajectory of the particles, after leaving the streamline, is tangential to the collecting body. Evidently, particles traveling on streamlines closer to the x axis than the limiting streamline will all be collected, and particles traveling outside this streamline will not be collected at all. Therefore, the collection efficiency of the collecting body can be expressed as

$$\eta_I = y_{\mathrm{lim}}/(D/2) = \tilde{y}_{\mathrm{lim}} \qquad \text{(cylindrical collector)} \tag{4.14a}$$

$$\eta_I = [y_{\mathrm{lim}}/(D/2)]^2 = \tilde{y}^2_{\mathrm{lim}} \qquad \text{(spherical collector)} \tag{4.14b}$$

Here, η_I gives the fractional number of those particles captured by the collector that were originally traveling upstream of the collector on streamlines ranging from $y = 0$ to $y = D/2$.

The first important investigation of inertial impaction was done by Sell (1931), who determined velocity profiles experimentally by studying streamlines in water moving past bodies of various shapes, such as a sphere, cylinder, and flat plate. Using the experimental streamlines, Sell calculated particle trajectories assuming that the particles had mass but negligible size. Albrecht (1931) calculated the limiting trajectories of particles using the potential flow model. Langmuir and Blodgett (1944) and Bosanquet (1950) also used the potential flow theory to determine limiting particle trajectories.

Measured inertial impaction efficiencies obtained by Landahl and Hermann (1949) for droplets *on cylinders* did not agree with Sell's predicted values; therefore, an efficiency curve was derived from fluid velocities at $\mathrm{Re_c} = 10$ obtained by Thom (1933). This empirical relationship describes their results (Pich, 1966):

$$\eta_\mathrm{I} = \frac{\Psi^3}{\Psi^3 + 0.77\Psi^2 + 0.22} \tag{4.15}$$

Extensive measurements of collection efficiencies of drops on cylinders and spheres have been made by Wong and Johnstone (1953), Ranz and Wong (1952), and Ranz *et al.* (1955). Friedlander and Pasceri (1967) correlated the experimental data of Wong and Johnstone (1953) by

$$\eta_\mathrm{I} = 0.075\Psi^{6/5} \tag{4.16}$$

for $0.8 < \Psi < 2$, $\mathrm{Re_c} < 1$ and $R < 0.2$.

Subramanyam and Kuloor (1969), assuming potential flow and point particles, calculated η_I in the range $\Psi < 20$:

$$\eta_\mathrm{I} = \frac{\Psi}{\Psi + 1.15} \tag{4.17}$$

The various calculated and experimental efficiency curves have been plotted versus the square root of the inertial impaction parameter on Fig. 4.3. The agreement between the various curves cannot be regarded as satisfactory. The curve at the bottom of the graph was calculated by Davies (1950) in the viscous flow region, where Stokes' law is valid. It appears that under viscous flow conditions the efficiency of inertial impaction is far lower than at high collector Reynolds number. The experimental results obtained by Wong and Johnstone (1953) are shown in Fig. 4.4. According to theory, there is a value, Ψ_crit, below which $\eta_\mathrm{I} = 0$. In potential flow, Ψ_crit is 0.0625 and 0.0417 for cylinders and for spheres, respectively (Langmuir and Blodgett, 1944, Bonsanquet, 1950). In reality, there are usually other mechanisms of deposition involved, so that even for $\Psi < \Psi_\mathrm{crit}$, the value of η is greater than 0.

The experimental results for inertial impaction by spheres on spheres obtained by Ranz and Wong (1952) and Jarman (1959) are shown in Fig. 4.5, as is the theoretical curve calculated by Langmuir and Blodgett (1944) for potential flow. Comparing the curves in Figs. 4.3, 4.4, and 4.5, one can see that for the same value of the inertial impaction parameter, the collection efficiency is considerably higher in the case of spheres than of cylinders.

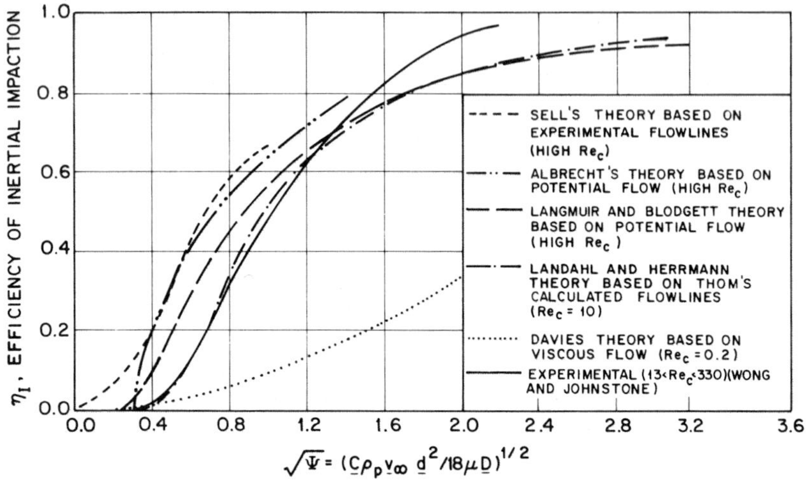

Figure 4.3 Efficiency of inertial impaction by spheres on cylinders based on various theories and the experiments by Wong and Johnstone. [Reprinted with permission from Strauss (1975).]

Figure 4.4 Experimental results of inertial impaction experiments by spheres impacting on wires (Wong and Johnstone, 1953).

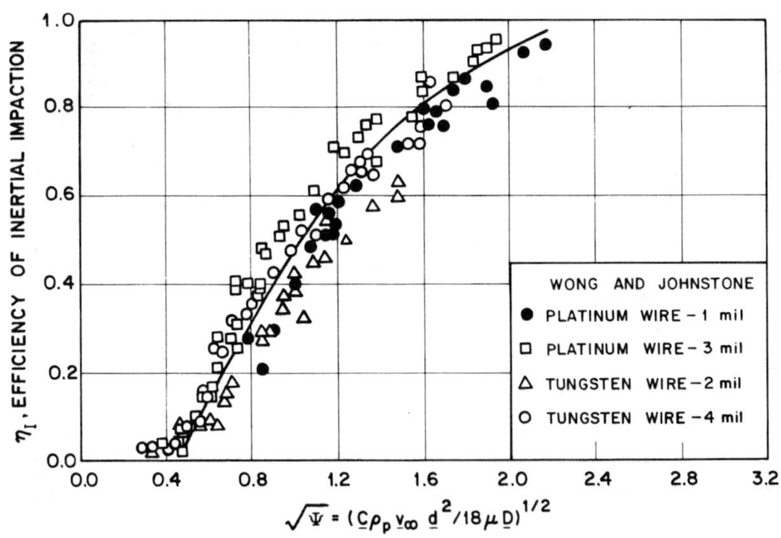

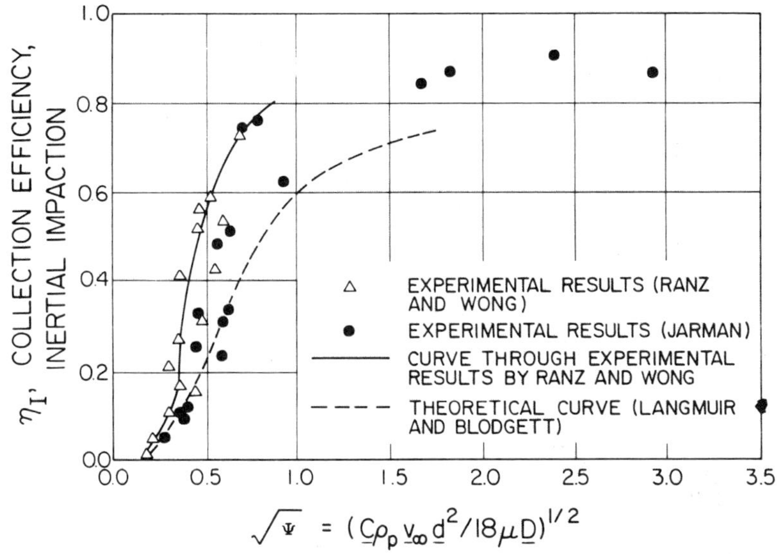

Figure 4.5 Experimental results for inertial impaction by spheres on spheres, by Ranz and Wong (1952) and Jarman (1959), together with the theoretical curve by Langmuir and Blodgett (1944).

Two empirical formulas have been obtained by Langmuir and Blodgett (1946) to give the collections efficiency by spherical collectors (1) at high collector Reynolds numbers ($Re_c \geq 10$):

$$\eta_I = \frac{(2\Psi)^2}{(2\Psi + 0.25)^2} \qquad (\Psi > 0.3) \tag{4.18}$$

and (2) in the viscous flow region:

$$\eta_I = \left[1 + \frac{0.75 \ln(4\Psi)}{2\Psi - 1.214}\right]^{-2} \qquad (2\Psi_{crit} = 1.214) \tag{4.19}$$

As mentioned, the turbulence intensity of the gas stream carrying aerosol particles can also be expected to have an effect on the collection efficiency, because the particles undergo eddy diffusion of varying intensity, depending on the degree of turbulence of the gas stream. Eddy diffusion, much like molecular diffusion, can be expected to result in collection, provided that the particles carried by the eddies can penetrate the boundary layer and reach the collector surface.

4.3 Interception

The Interception Model

The finite size of the dust particles has been ignored in the previous section on the analysis of inertial impaction of the particles on a collector body. The effect of the particle size on the collection efficiency is considered in a separate calculation in which the particles are assumed to have no mass but a finite size. Evidently, such particles will always follow the streamlines, because, having no mass, they have no inertia either. According to this mechanism, the particle gets collected when, while traveling along a streamline, it gets close enough to the collecting body to actually touch it. This will happen when the streamline on which the particle's center approaches the collector to a distance equal to $d/2$, as shown in Fig. 4.2.

Collection Efficiency by Interception Mechanism

The efficiency of particle collection by interception mechanism depends only on the ratio $R = d/D$, if potential flow is assumed. These efficiency formulas have been derived for cylindrical and spherical collectors, respectively (Ranz and Wong, 1952):

$$\eta_C = (1 + R) - \frac{1}{1 + R} \tag{4.20}$$

and

$$\eta_C = (1 + R)^2 - \frac{1}{1 + R} \tag{4.21}$$

For the case of the viscous flow model, the collection efficiency is influenced by the collector Reynolds number, $\mathrm{Re_c}$, in addition to the parameter R. For cylindrical collectors (Ranz, 1953),

$$\eta_C = \frac{1}{2 - \ln \mathrm{Re_c}} \left[(1 + R)\ln(1 + R) - \frac{R(2 + R)}{2(1 + R)} \right] \tag{4.22}$$

It is evident from the definition of the collection efficiency by interception that η_C can be greater than 1. The physical meaning of this is that such particles whose centers travel on streamlines farther away from the x axis than the distance $D/2$ may be collected by interception mechanism, provided that their diameter is large enough. Although it may be disturbing to see efficiencies in excess of unity, the formulas are consistent and can be used without introducing any error. One only has to bear in mind the fact that the efficiency η is based on particles whose centers travel, far upstream from the collecting body, on streamlines such that $\tilde{y}$ lies between -1 and $+1$.

For small values of R ($R \ll 1$), Eq. (4.22) simplifies to

$$\eta_C = \frac{R^2}{2 - \ln \mathrm{Re_c}} \tag{4.23}$$

As discussed, in reality the mechanisms of particle collection by inertial impaction and interception are not independent of each other, as was assumed in most calculations. Ideally, the calculation should have been carried out by taking into consideration both effects simultaneously. This has been done by Davies (1952) in the viscous flow region for cylindrical collectors ($\mathrm{Re_c} = 0.2$). His results are expressed analytically as

$$\eta_{IC} = 0.16[R + (0.50 + 0.8R)\Psi - 0.1052R\Psi^2] \tag{4.24a}$$

Note also that (see Example 4.4) Ψ is related to R as

$$\Psi = \frac{C\rho_p R^2 \mathrm{Re_c}}{18\rho} \tag{4.24b}$$

As in any real particle collection problem ρ, ρ_p, and C have fixed values, Ψ is not independent of R; therefore, Eqs. (4.24a) and (4.24b) must be used jointly.

Often, it is assumed that the two mechanisms of particle collection are statistically independent of each other and, with this assumption, the combined collection efficiency is calculated as

$$\eta_{IC} = 1 - (1 - \eta_C)(1 - \eta_I) \tag{4.25}$$

Equation (4.25) is based on the theory of probability of independent events. Collection by inertial impaction and collection by interception are regarded as two independent events. Then one can write: Start with 1 unit of aerosol by mass. After inertial impaction, $(1 - \eta_I)$ units are left. Of this, $\eta_C(1 - \eta_I)$ are collected by interception, and

$$(1 - \eta_I) - \eta_C(1 - \eta_I) = (1 - \eta_C)(1 - \eta_I)$$

are left uncollected. Thus, the overall efficiency,

$$\eta_{IC} = \frac{\text{total collected amount}}{\text{initial amount}} = \frac{\text{initial amount} - \text{uncollected amount}}{\text{initial amount}}$$

$$= \frac{1 - (1 - \eta_C)(1 - \eta_I)}{1} = 1 - (1 - \eta_C)(1 - \eta_I)$$

EXAMPLE 4.2

Calculate the collection efficiency by inertial impaction in potential flow and in viscous flow for $\Psi = 0.1$ and $\Psi = 2.0$.

SOLUTION

1. Potential flow, $\Psi = 0.1$:
   a. Spherical collector:

$$\eta_{\mathrm{I}} = \frac{(2\Psi)^2}{(2\Psi^2 + 0.25)^2} = 0.1975 \tag{4.18}$$

   b. Cylindrical collector:

$$\eta_{\mathrm{I}} = \frac{\Psi^3}{\Psi^3 + 0.77\Psi^2 + 0.22} = 0.0044 \tag{4.15}$$

$$\eta_{\mathrm{I}} = \frac{\Psi}{\Psi + 1.15} = 0.08 \tag{4.17}$$

   Potential flow, $\Psi = 0.20$:
   a. Spherical collector:

$$\eta_{\mathrm{I}} = 0.886$$

   b. Cylindrical collector:

$$\eta_{\mathrm{I}} = 0.708, 0.635$$

2. Viscous flow, $\Psi = 0.1$:
   a. Spherical collector:

$$\eta_{\mathrm{I}} = \left[1 + \frac{0.75\ln(4\Psi)}{2\Psi - 1.214}\right]^{-2} = 0 \tag{4.19}$$

   b. Cylindrical collector:

$$\eta_{\mathrm{I}} = 0.075\Psi^{6/5} = 0.005 \tag{4.16}$$

   Viscous flow, $\Psi = 2.0$:
   a. Spherical collector:

$$\eta_{\mathrm{I}} = 0.411$$

   b. Cylindrical collector:

$$\eta_{\mathrm{I}} = 0.172$$

EXAMPLE 4.3

Calculate collection efficiencies by interception in potential flow and in viscous flow as a function of the parameter $R = d/D$.

SOLUTION

	η_C			
	Potential flow		Viscous flow	
R	Cylinder Eq. (4.20)	Sphere Eq. (4.21)	Cylinder Eq. (4.24a) $\Psi = 0$	Cylinder Eq. (4.22)
0.01	0.02	0.03	0.0016	0.0000275
0.10	0.19	0.30	0.016	0.0026
0.33	0.578	1.00	0.053	0.025
0.5	0.833	1.58	0.086	0.053
0.6	0.975	1.94	0.096	0.073
0.618	1.00	2.00	0.099	0.096
1.00	1.50	3.50	0.16	0.176

Note these points:

1. η_C is much larger in potential flow than in viscous flow.
2. η_C is larger for spheres than for cylinders.
3. η_C may be greater than unity. This is simply a consequence of the definition used in calculating η_C.

EXAMPLE 4.4

Are $R = d/D$, Ψ, and Re_c all independent? Derive a relationship between R, Ψ, and Re_c. Then, using this relationship, calculate and tabulate η_I, η_C, and η_{IC} for cylindrical and spherical collectors as functions of R, for $C = 1$, $\rho_p = 10^3$ kg/m^3, and $\rho = 1$ kg/m^3 in viscous flow ($\text{Re}_c = 0.2$) and potential flow ($\text{Re}_c = 100$).

SOLUTION

As $R = d/D$, it follows that $d^2 = R^2D^2$. Substituting R^2D^2 for d^2 in the relationship $\Psi = C\rho_p d^2 v_\infty/18\mu D$, one obtains

$$\Psi = \frac{C\rho_p R^2 D^2 v_\infty}{18\mu D} = \frac{C\rho_p R^2 D v_\infty}{18\mu}$$

But, from $\text{Re}_c = Dv_\infty\rho/\mu$, it follows that

$$\frac{Dv_\infty}{\mu} = \frac{\text{Re}_c}{\rho}$$

whence there is this relationship between Ψ, R, and Re_c:

$$\Psi/(R^2\,\text{Re}_c) = C\rho_p/18\rho$$

Thus, for fixed values of C (i.e., d/λ), ρ, and ρ_p, the group $\Psi/(R^2\,\text{Re}_c)$ is a constant.

For viscous flow ($Re_c = 0.2$) $\Psi = 11.1R^2$.

R	Ψ	η_I (sphere) Eq. (4.19)	η_C (cylinder) Eq. (4.22)	η_{IC} (cylinder) Eq. (4.24a)
0.01	0.0011	0	0.00003	0.0017
0.1	0.011	0	0.0026	0.026
0.33	1.21	0.25	0.025	0.193
0.5	2.77	0.50	0.053	0.415
0.6	4.00	0.59	0.073	0.562
0.7	5.44	0.65	0.096	0.687

For potential flow ($Re_c = 100$) $\Psi = 5.56 \times 10^3 R^2$.

R	Ψ	η_I (sphere) Eq. (4.18)	η_C (sphere) Eq. (4.21)	η_I (cylinder) Eq. (4.17)	η_C (cylinder) Eq. (4.20)
0.01	0.556	0.67	0.03	0.27	0.02
0.1	55.6	0.996	0.30	0.99	0.19
0.33	—	1.00	1.00	1.00	0.578
0.5	—	1.00	1.58	1.00	0.833
0.6	—	1.00	1.94	1.00	0.975
0.618	—	1.00	2.00	1.00	1.00
1.00	—	1.00	3.50	1.00	1.50

4.4 Particle Collection by Brownian Diffusion

For particle sizes below 1 μm, the contribution of the inertial impaction mechanism to particle collection gradually decreases and collection by Brownian diffusion becomes more significant. For particle sizes under 0.1 μm, Brownian diffusion is the dominant mechanism of particle collection.

Relatively large particles do not show any detectable Brownian motion, because the number of gas molecules impinging on their surface from opposing directions is very nearly the same at any moment. With decreasing particle size, however, the rate of bombardment of the particles by gas molecules becomes less and less uniform and, as a result, a net force of random magnitude and direction acts on the particle at any moment. Owing to this force, the particle undergoes random motion, commonly referred to as *random walk*. Brownian motion of particles is random walk motion. If a large number of particles are undergoing random walk, the system of particles is said to be undergoing diffusion. An essential feature of random walk motion is that the magnitude and the direction of each step taken by the particle is independent of the direction and magnitude of the previous steps. Brownian motion can actually be observed under a micro-

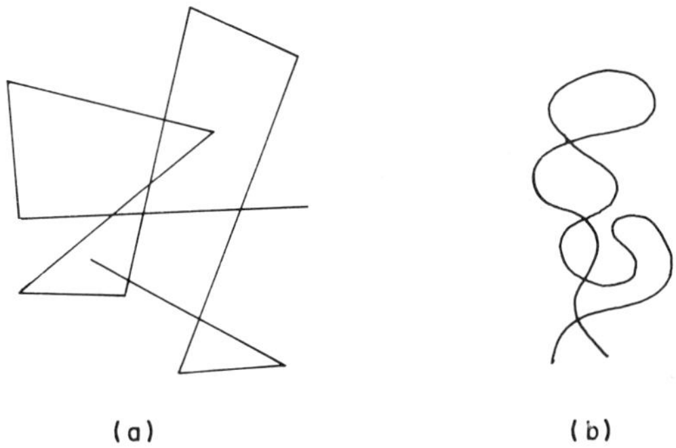

Figure 4.6 Trajectories of gas molecules (a) and particles undergoing Brownian motion (b). [Reprinted with permission from N. A. Fuchs, "Mechanics of Aerosols," Copyright (1964), Pergamon Books Ltd.]

scope of sufficient magnification, and a schematic diagram showing Brownian motion of a particle is shown in Fig. 4.6.

Estimation of the Particle Diffusion Coefficient

Einstein (1908) showed that the mean square displacement $\overline{x^2}$ of a particle in Brownian motion in time t is

$$\overline{x^2} = 2\mathscr{D}t \tag{4.26}$$

where

$$\mathscr{D} = kTB \tag{4.27}$$

is the diffusion coefficient of the particle, k is the Boltzmann constant, T is the absolute temperature, and B is the particle mobility defined as

$$B = \frac{\text{velocity of particle, } u}{\text{resisting force causing steady motion, } F} \tag{4.28}$$

For example, when Stokes' law is valid the resisting force causing steady motion is given by Eq. (2.12). Hence, applying the definition expressed in Eq. (4.28), we have

$$B = u/F = 1/3\pi\mu d \tag{4.29}$$

Generally, the Cunningham correction factor must be used, and the expression for the mobility becomes

$$B = C/3\pi\mu d \tag{4.30}$$

It is apparent from these equations that the particle mobility is determined by the size of the particle, the viscosity of the medium in which the particle is suspended, and the mean free path of the molecules of the medium. According to Eq. (4.27), which is valid for particle sizes of an order equal to or greater than the mean free path, the diffusion coefficient is proportional to the mobility of the particles and to the absolute temperature.

When the particles are much smaller than the mean free path, they can be treated as large molecules. The formula derived in the kinetic theory of gases for the molecular diffusion coefficient has been modified by Langmuir (1942) to calculate the diffusion coefficient of aerosol particles as

$$\mathscr{D} = (4kT/3\pi d^2 p)(8RT/\pi M)^{1/2} \tag{4.31}$$

where p is the hydrostatic pressure and M is the molecular weight of the gas. The two predictive equations of the diffusion coefficient are compared in Example (4.5).

Diffusional Deposition of Aerosol Particles from a Stationary Gas and from Laminar Flow

In a stationary gas, the mass balance combined with Fick's law of diffusion gives this differential equation of diffusion:

$$\partial N/\partial t = \mathscr{D}\nabla^2 N \tag{4.32}$$

where N is the number of particles per unit volume of gas. In a gas flowing with a velocity $\mathbf{v}$, however, there is also a convective term in the differential equation:

$$\partial N/\partial t = -\mathrm{div}(\mathbf{v}N) + \mathscr{D}\nabla^2 N \tag{4.33}$$

In spherical polar coordinates Eq. (4.32) becomes

$$\frac{\partial (Nr)}{\partial t} = \mathscr{D}\frac{\partial^2 (Nr)}{\partial r^2} \tag{4.34}$$

which has been solved analytically for the case of various initial and boundary conditions. The rate of deposition Φ on the surface of a spherical collector obtained from the solution of this equation is

$$\Phi = 2\pi D\mathscr{D}N_0\left(1 + \frac{D}{2\sqrt{\pi\mathscr{D}t}}\right) \tag{4.35}$$

Here, N_0 is the initial number of particles per unit volume of gas. It can be seen from Eq. (4.35) that when $\mathscr{D}t/D^2 \gg 1$ the time dependence disappears, that is, steady state is reached, and the rate of deposition becomes:

$$\Phi = 2\pi D\mathscr{D}N_0 \tag{4.36}$$

The problem of calculating the rate of deposition of the particles on spheres or cylinders by diffusion from laminar flow is much more difficult and it has not been done rigorously. A large number of relationships for the rate of deposition of particles by diffusion from laminar flow have been obtained, and most are represented here in the form of a particle collection efficiency, $\eta_{\mathscr{D}}$, which is defined analogously as the efficiencies of particle collection by inertial impaction and by interception. According to this definition there follows that the particle collection efficiency by diffusion mechanism from a stationary gas is always infinite because, in this case, the number of particles carried by the gas stream across an area equal to the normal projection of the collector is equal to zero for any time t. Therefore, it is not surprising that we often obtain values for $\eta_{\mathscr{D}}$ in excess of unity whenever the gas flow is slow.

In these correlations, a dimensionless parameter, the Schmidt number Sc

$$\mathrm{Sc} = \nu/\mathscr{D} \tag{4.37}$$

is used, in addition to the collector Reynolds number, $\mathrm{Re_c}$. (ν is kinematic viscosity.)

Instead of the Schmidt number, the diffusional variant of the Peclet number, known from heat transfer, is often used:

$$\mathrm{Pe} = \mathrm{Re_c}\,\mathrm{Sc} = \frac{v_\infty \rho D}{\mu}\,\frac{\mu}{\rho \mathscr{D}} = \frac{v_\infty D}{\mathscr{D}} \tag{4.38}$$

In isothermal mass transfer, the Schmidt number plays a role analogous to the role played by the Prandtl number in heat transfer.

Diffusional Collection Efficiency of a Cylindrical Collector

Three regions as defined below may be distinguished:

1. $\mathrm{Pe} \ll 1$, $\mathrm{Re_c} < 1$:

$$\eta_{\mathscr{D}} = \frac{2\pi}{\mathrm{Pe}(1.502 - \ln \mathrm{Pe})} \tag{4.39}$$

(Stechkina, 1964).

2. $\mathrm{Pe} \gg 1$, $\mathrm{Re_c} < 1$:

$$\eta_{\mathscr{D}} = \frac{1}{2(2 - \ln \mathrm{Re_c})}\left[2(1 + Z)\ln(1 + Z) - (1 + Z) + \frac{1}{1 + Z}\right] \tag{4.40}$$

where

$$Z = 1.308\left[(2 - \ln \mathrm{Re_c})^{1/3}/\mathrm{Pe}^{1/3}\right] \tag{4.41}$$

(Langmuir, 1942).

For large Pe, that is, for $Z \ll 1$, this equation reduces to

$$\eta_{\mathscr{D}} = 1.71 \frac{1}{(2 - \ln \mathrm{Re_c})^{1/3}} \mathrm{Pe}^{-2/3} \tag{4.42}$$

Friedlander (1957) and Natanson (1957) have also obtained the same expression, but with a coefficients of 2.22 and 2.92, respectively.

Johnstone and Roberts (1949) and Ranz (1951) applied the analogy between heat and mass transfer and obtained

$$\eta_{\mathscr{D}} = \frac{1}{\mathrm{Pe}} + 1.727 \frac{\mathrm{Re_c^{1/6}}}{\mathrm{Pe}^{2/3}} \tag{4.43}$$

Torgeson (1958) obtained the expression:

$$\eta_{\mathscr{D}} = 0.75 \left[\frac{4\pi}{2 - \ln \mathrm{Re_c}} \right]^{0.4} \mathrm{Pe}^{-0.6} \tag{4.44}$$

The equation of Stechkina (1966) is valid for $\mathrm{Pe} > 80(2 - \ln \mathrm{Re_c})$:

$$\eta_{\mathscr{D}} = \frac{2.9}{(2 - \ln \mathrm{Re_c})^{1/3}} \mathrm{Pe}^{-2/3} + 0.624\, \mathrm{Pe}^{-1} \tag{4.45}$$

3. $\mathrm{Pe} \gg 1$, $\mathrm{Re} \gg 1$:

$$\eta_{\mathscr{D}} = C \frac{1}{\mathrm{Pe}^{1/2}} \tag{4.46}$$

where $C = 2.83$ (Bosanquet, 1950), $C = 3.19$ (Boussinesq, 1903, Natanson, 1957), and $C = \pi/2$ (Lewis and Smith, 1942).

Combined Collection Efficiency of a Cylindrical Collector

These equations have been obtained for the combined collection efficiency by diffusion and interception:

$$\eta_{\mathscr{D}C} = \frac{1}{2 - \ln \mathrm{Re_c}} \left[(1 + 2x) \ln(1 + 2x) - \frac{x(2 + 2x)}{1 + 2x} \right] \tag{4.47}$$

where x is the solution of the equation

$$x\left[x - \tfrac{1}{2}R\right]^2 = 0.28(2 - \ln \mathrm{Re_c})\, \mathrm{Pe}^{-1} \tag{4.48}$$

(Langmuir, 1942). Friedlander (1958) derived the semiempirical equation

$$\eta_{\mathscr{D}C} = \frac{6\, \mathrm{Re_c^{1/6}}}{\mathrm{Pe}^{2/3}} + 3R^2\, \mathrm{Re_c^{1/2}} \tag{4.49}$$

Stechkina and Fuchs (1966) derived the equation

$$\eta_{\mathscr{D}C} = \eta_{\mathscr{D}} + \eta_C + 1.24 \frac{R^{2/3}}{(2 - \ln \mathrm{Re_c})^{1/2}\, \mathrm{Pe}^{1/2}} \tag{4.50}$$

where $\eta_{\mathscr{D}}$ is given by Eq. (4.45) and η_C by Eq. (4.22).

For $\mathrm{Re_c} < 1$, $R \ll 1$, and $\mathrm{Pe} \gg 1$, Radushkevich and Kolganov (1969) derived the next expression, representing an approximate solution of a more general formulation given by Friedlander (1967).

$$\eta_{\mathscr{D}C} = \frac{4.35}{2^{1/3}} \frac{1}{(2 - \ln \mathrm{Re_c})^{1/3}} \frac{\mathrm{Pe}^{-2/3}}{0.893 - x} \tag{4.51a}$$

where

$$x = \frac{R^3\, \mathrm{Pe}}{6(2 - \ln \mathrm{Re})} \tag{4.51b}$$

This empirical correlation is from Aiba and Yasuda (1962):

$$\eta_{\mathscr{D}C} = \mathrm{Re_c}^{1/18}\, \mathrm{Pe}^{-2/3} + R^2\, \mathrm{Re_c}^{1/6} \tag{4.52}$$

Another empirical relation, from Friedlander and Pasceri (1967), has been written in the form

$$\eta_{\mathscr{D}C} = 1.3\, \mathrm{Pe}^{-2/3} + 0.7R^2 \tag{4.53}$$

A number of relations are also available for the combined collection efficiency by diffusion, interception, and inertial impaction. The semiempirical equation from Davies (1952) for $\mathrm{Re_c} = 0.2$ is presented here:

$$\eta_{\mathscr{D}CI} = 0.16\left[R + (0.5 + 0.8R)(\mathrm{Pe}^{-1} + \Psi) - 0.105R(\mathrm{Pe}^{-1} + \Psi)^2\right] \tag{4.54}$$

Diffusional Collection Efficiency of a Spherical Collector

The equations included here are

$$\eta_{\mathscr{D}} = (8/3\pi)(2/\mathrm{Pe})^{1/2} \tag{4.55}$$

(Strauss, 1959);

$$\eta_{\mathscr{D}} = (4/\mathrm{Pe})(2 + 0.557\, \mathrm{Re_c}^{1/2}\, \mathrm{Sc}^{3/8}) \tag{4.56}$$

(Johnstone and Roberts, 1949);

$$\eta_{\mathscr{D}} = \frac{\pi \mathscr{D} D\, \mathrm{Re_c}^{1/3}\, \mathrm{Sc}^{1/3}}{\tilde{Q}}; \qquad \mathrm{Re_c} < 3; \qquad \mathrm{Sc} \approx 10^6 \tag{4.57}$$

where $\tilde{Q}$ is the gas flow rate in a stream tube having a cross section equal to that of the collector (Levich, 1962);

$$\eta_{\mathscr{D}} = \frac{0.8\pi\mathscr{D}D\,\mathrm{Re}_c^{1/2}\,\mathrm{Sc}^{1/3}}{\tilde{Q}}; \qquad 600 < \mathrm{Re}_c < 2600; \qquad \mathrm{Sc} \approx 10^6 \quad (4.58)$$

(Akselrud, 1953); and

$$\eta_{\mathscr{D}} = \frac{0.95\pi\mathscr{D}D\,\mathrm{Re}_c^{1/2}\,\mathrm{Sc}^{1/3}}{\tilde{Q}}; \qquad 100 < \mathrm{Re}_c < 700; \qquad \mathrm{Sc} \approx 10^3 \quad (4.59)$$

(Garner and Keey, 1958; Garner and Suckling, 1958).

Combined Collection Efficiency of a Spherical Collector

These equations have been developed for the combined collection efficiency of spherical collectors:

$$\eta_{\mathscr{D}C} = 2.18\,\mathrm{Pe}^{-1/2} + 2(d/D); \qquad d/D \le 0.3\,\mathrm{Pe}^{-1/2} \quad (4.60)$$

(Rimberg and Peng, 1977);

$$\eta_{\mathscr{D}CI} = 1.53(d/2)^{-0.85}10^{-[9.9(D/2)+5]}; \qquad 100 < \mathrm{Re}_c < 5000 \quad (4.61)$$

(Rimberg and Peng, 1977); and

$$\eta_{\mathscr{D}CI} = 3.02\,\mathrm{Re}_c^{1/6}\,\mathrm{Pe}^{-2/3} + 1.14(d/D)\,\mathrm{Re}^{1/3}\,\mathrm{Pe}^{-1/3} + 0.57(d/D)^2\,\mathrm{Re}_c^{1/3}; \qquad \mathrm{Re}_c > 500 \quad (4.62)$$

(Rimberg and Peng, 1977).

EXAMPLE 4.5

Calculate the Brownian diffusion coefficient, $\mathscr{D}$, of particles of diameter $d = 10^{-2}$ μm in air at 20°C and 1 atm pressure.

SOLUTION

1. Using Einstein's equation:

$$\mathscr{D} = kTC/3\pi\mu d \qquad \text{[Eqs. (4.27) and (4.30)]}$$

$$k = \frac{R}{N} = \frac{8.3144\ \mathrm{J/g\ mol\ K}}{6.02 \times 10^{23}\ \mathrm{molecule/g\ mol}} = \frac{1.38 \times 10^{-23}\ \mathrm{J}}{(\mathrm{K} \times \mathrm{molecule})}$$

$$C = 1 + (2\lambda/d)[A + Q\exp(-bd/2\lambda)] \qquad \text{[Eq. (2.46)]}$$

where

$$A = 1.246, \qquad Q = 0.42, \qquad b = 0.87$$

$$\therefore \lambda = \frac{\mu}{0.5\rho(8RT/\pi M)^{1/2}} = \frac{1.82 \times 10^{-5}}{(0.5)(1.2)(8 \times 8.3144 \times 10^3 \times 293/29\pi)^{1/2}}$$

$$= 6.56 \times 10^{-8}\ \text{m} \qquad [\text{Eqs. (2.41) and (2.42)}]$$

$$\therefore C = 1 + \frac{(2)(6.56 \times 10^{-8})}{10^{-8}}$$

$$\times \{1.246 + 0.42 \exp[-0.87 \times 10^{-8}/(2 \times 6.56 \times 10^{-8})]\} = 22.5$$

$$\therefore \mathscr{D} = \frac{(1.38 \times 10^{-23})(293)(22.5)}{3\pi(1.82 \times 10^{-5})(10^{-8})} = 5.3 \times 10^{-8}\ \text{m}^2\ \text{s}^{-1}$$

2. Using Langmuir's equation:

$$\mathscr{D} = \frac{4kT}{3\pi d^2 p}\left(\frac{8RT}{\pi M}\right)^{1/2} \tag{4.31}$$

$$p = 1\ \text{atm} = 98.07 \times 10^3\ \text{Pa}$$

$$\mathscr{D} = \frac{(4)(1.38 \times 10^{-23})(293)}{3\pi(10^{-16})(98.07 \times 10^3)}\left[\frac{(8)(8.3144 \times 10^3)(293)}{29\pi}\right]^{1/2}$$

$$= 8.1 \times 10^{-8}\ \text{m}^2\ \text{s}^{-1}$$

Values of $\mathscr{D}$ calculated by these equations for different particle diameters are tabulated here:

	$\mathscr{D}$ ($m^2\ s^{-1}$)	
d (μm)	Einstein Eq. (4.27)	Langmuir Eq. (4.31)
10	2.4×10^{-12}	8.1×10^{-14}
1	2.7×10^{-11}	8.1×10^{-12}
0.1	6.1×10^{-10}	8.1×10^{-10}
0.01	5.3×10^{-8}	8.1×10^{-8}
0.001	4.0×10^{-6}	8.1×10^{-6}

It is apparent that the two equations yield consistent results in the neighborhood of $d = 0.5\ \mu$m. Above this value, the Einstein formula is to be preferred.

EXAMPLE 4.6

Calculate and tabulate the values of $\eta_{\mathscr{D}}$ for a cylindrical collector of $D = 1$ μm in air at $t = 20°$C and 1 atm pressure for $v_\infty = 0.1$ m s^{-1}, as a function of the particle diameter d. Use all the equations that are applicable.

SOLUTION

		$\eta_{\mathscr{D}}$ ($Re_c = 6.7 \times 10^{-3}$)					
d (μm)	Pe	Experimental	Eq. (4.39)	Eq. (4.40)	Eq. (4.43)	Eq. (4.44)	Eq. (4.45)
10	4.2×10^4		N.A.	7.4×10^{-4}	6.4×10^{-4}	1.6×10^{-3}	1.2×10^{-3}
1	3.7×10^3		N.A.	3.7×10^{-3}	3.4×10^{-3}	6.8×10^{-3}	6.4×10^{-3}
0.1	1.63×10^2	0.18	N.A.	3.0×10^{-2}	3.1×10^{-2}	4.4×10^{-2}	N.A.
0.01	1.9		N.A.	N.A.	N.A.	N.A.	N.A.
0.001	0.025		48	N.A.	N.A.	N.A.	N.A.

It is apparent that the different equations predict different values for $\eta_{\mathscr{D}}$ and that the predictions can be considered order of magnitude estimates, at best. A dearth of experimental data makes it very difficult to have a preference for any particular equation.

PROBLEMS

1. Estimate the combined collection efficiency $\eta_{\mathscr{D}C}$ of a single (cylindrical) fiber 1 μm in diameter when dusty air containing 0.1-μm dust particles is flowing past the fiber (perpendicular to the fiber axis) at a velocity of 10 cm s^{-1}. Assume ambient conditions. Also, calculate the inertia parameter to verify that under these conditions collection by an inertial impaction mechanism is negligible. (Assume a particle density of 3×10^3 kg m^{-3}.)

2. Consider the flow through a fiber filter. The average diameter of the cylindrical fibers is 100 μm. Other data are

$$\rho_p = 2.58 \text{ g/cm}^3$$

$$\rho = 1.2 \times 10^{-3} \text{ g/cm}^3$$

$$\mu = 1.8 \times 10^{-4} \text{ g/cm}$$

The particle size distribution shows that the particle diameter ranges from 0.001 μm to 1.0 μm. The velocity in the filter varies from 200 to 1000 cm/s. Calculate the inertial impaction target collection efficiency of a single cylindrical fiber as a function of the particle diameter for velocities of 200 and 1000 cm/s. The collector Reynolds number is given by

$$Re_c = Dv_\infty \rho/\mu$$

In this problem the collector Reynolds number varies from 13.3 to 66.7. Since Re_c is greater than 10, the flow pattern around the collector may be considered to be potential flow.

3. Consider problem 2. Calculate the target collection efficiency by interception as a function of the particle diameter for velocities of 200 and 1000 cm/s.

4. Consider problems 2 and 3. Assuming ambient conditions (T = 20°C and P = 101 kPa), predict the collection efficiency of diffusion as a function of aerosol diameter for velocities of 200 and 1000 cm/s.

5. Considering problems 2–4, estimate the overall target efficiency of the cylinder, taking into account inertial impaction, interception, and diffusion. Calculate the collection efficiency as a function of the particle diameter for velocities of 200 and 1000 cm/s.

6. Dusty gas flow in a duct is cleaned according to a new invention by placing a stack of parallel screens in the duct. The dust deposits on the wires or fibers of the screens, and the dust buildup falls off the screen in large chunks, which are collected in hoppers, much like in a gravity settler. The basic idea of this device is to capture and agglomerate the dust particles on screens.

   a. Estimate and tabulate the collection efficiency in a 2-ft-diameter duct for 2 m s^{-1} and 6 m s^{-1} air velocities, and 0.05-, 0.2-, 0.5-, 1.0-, 5.0-, 10-, and 30-μm dust particles in ambient air with these dimensions:
      i. $\frac{1}{4}$ inch mesh; wire diameter, 0.6 mm
      ii. $\frac{1}{8}$inch mesh; wire diameter, 0.3 mm

   b. Estimate the efficiency of a stack of
      i. 10 screens
      ii. 100 screens
      under the conditions of part a, assuming that all the particles impacting on the screens are captured and ultimately collected. That is, assume no bouncing off of individual particles or reentrainment of the dust already deposited on the screens. Discuss your results.

7. Repeat the calculations of problem 6, using the same duct diameter, dust particle diameters, and ambient conditions, but air velocities of 2 cm s^{-1}, 50 cm s^{-1}, and 2 m s^{-1}, and screens with these dimensions:

   a. 1 mm mesh; fiber diameter, 50 μm

   b. 0.1 mm mesh; fiber diameter, 5 μm
   Discuss your results.

8. a. Water drops 1 mm in diameter are settling in dusty air at 1 atm pressure and 50°C. The air is streaming upward, countercurrent to the drops. Estimate the collection efficiency of 1 drop for these dust

particle diameters; 100, 50, 30, 10, 1, and 0.1 μm. The dust particle density is 5.5 g cm^{-3}.

b. Repeat the calculations of part a for water drops with a diameter of 1 mm. Discuss your results.

References

Aiba, S., and Yasuda, T. (1962). A correlation between single fiber efficiencies of fibrous filters and operating variables. *AIChE J.* **8**, 704–707.

Akselrud, G. (1953). *Zh. Fiz. Khim.* (in Russian) **27**, 1445.

Albrecht, F. (1931). Theoretische Untersuchunger ueber die Ablagerung von Staub aus stroemender Luft und ihre Anwendung auf die Theorie der Staubfilter. *Phys. Z.* **32**, 48–56.

Bosanquet, C. H. (1950). *Trans. Inst. Chem. Eng. (London)* **28**, 130. (Appendix to paper by C. J. Stairmand.)

Boussinesq, J. (1903). "Théorie Analytique de Chaleur," Vol. II. Gauthier-Villars, Paris.

Davies, C. N. (1950). Viscous flow transverse to a circular cylinder. *Proc. Phys. Soc.* **63B**, 288–297.

Davies, C. N. (1952). Separation of airborne dust and particles. *Proc. Inst. Mech. Eng.* **1B**, 185–198.

Einstein, A. (1908). "Investigations on the Theory of Brownian Movement," p. 75. Dover, New York. [From *Z. Elektrochemie* **14**, 235.]

Friedlander, S. K. (1957). Mass and heat transfer to single spheres and cylinders at low Reynolds numbers. *AIChE J.* **3**, 43–48.

Friedlander, S. K. (1958). Theory of aerosol filtration. *Ind. Eng. Chem.* **50**, 1161–1164.

Friedlander, S. K. (1967). Particle diffusion in low-speed flows. *J. Colloid Interface Sci.* **23**, 157–16

Friedlander, S. K., and Pasceri, R. E. (1967). Aerosol filtration by fibrous filters. *In* "Biochemical and Biological Engineering" (N. Blakenbrough, ed.), pp. 50–67. Academic Press, New York.

Fuchs, N. A. (1964). *In* "Mechanics of Aerosols" (C. N. Davies, ed.), revised Ed. (trans.). Pergamon, Oxford.

Fuchs, N. A., and Stechkina, I. B. (1963). A note on the theory of fibrous aerosol filters. *Ann. Occup. Hyg.* **6**, 27–30.

Garner, F. H., and Keey, R. B. (1958). Mass transfer from single solid spheres, 1: Transfer at low Reynolds numbers. *Chem. Eng. Sci.* **9**, 119–129.

Garner, F. H., and Suckling, R. D. (1958). Mass transfer from a soluble solid sphere. *AIChE J.* **4**, 114–124.

Jarman, R. T. (1959). *J. Agric. Eng. Res.* **4**, 139.

Johnstone, H. F., and Roberts, M. H. (1949). Deposition of aerosol particles from moving gas streams. *Ind. Eng. Chem.* **41**, 2417–2423.

Landahl, H. D., and Hermann, R. G. (1949). Sampling of liquid aerosols by wires, cylinders, and slides, and the efficiency of impaction of the droplets. *J. Colloid Sci.* **4**, 103–136.

Langmuir, I. (1942). *OSRD* Rep. No. 865.

Langmuir, I., and Blodgett, K. (1944). Report R.L. 225. General Electric Research Laboratory, Schenectady, New York.

Langmuir, I., and Blogett, K. B. (1946). *Am. Air Force Tech. Rep.* 5418.

Levich, V. G. (1962). "Physicochemical Hydrodynamics" (English trans.). Prentice-Hall, Englewood Cliffs, New Jersy.

Lewis, W. K., and Smith, J. M. (1942). *OSRD Rep.* No. 1251.

Natanson, G. L. (1957). Diffusive deposition of aerosols on a cylinder in a flow in the case of small capture coefficients. *Dokl. Akad. Nauk SSSR* **112**, 100–103 (in Russian).

Pich, J. (1966). Theory of aerosol filtration. *In* "Aerosol Science" (C. N. Davies, ed.), Chap. 9. Academic Press, New York.

Radushkevich, L. V., and Kolganov, V. A. (1969). Experimental study of deposition of flowing highly dispersed aerosols in thin cylinders. *J. Colloid Interface Sci.* **29**, 55–65.

Ranz, W. E. (1951). *Tech. Rep.* No. 3. Univ. of Illinois, Urbana.

Ranz, W. E. (1953). *Tech. Report* No. 8. Univ. of Illinois, Urbana, Eng. Exp. Sta.

Ranz, W. E., and Wong, J. B. (1952). Impaction of dust and smoke particles on surface and body collectors. *Ind. Eng. Chem.* **44**, 1371–1381.

Ranz, W. E., Wong, J. B., and Johnstone, H. F. (1955). Inertial impaction of aerosol particles on cylinders. *J. Appl. Phys.* **26**, 244–249.

Rimberg, D., and Peng, Y.-M. (1977). Aerosol collection by falling droplets. *In* "Air Pollution Control and Design Handbook" (P. N. Cheremisinoff and R. A. Young eds.), Part 2. Dekker, New York.

Sell, W. (1931). Staubausscheidung an einfachen Koerpern und in Luftfiltern (Dust removal). *Ver. Deutsch. Ing. Forschungsh.* 347.

Stechkina, I. B. (1964). *Inzh. Fiz. Zh.* **7**, 128 (in Russian).

Stechkina, I. B. (1965). Diffusion precipitation of aerosols in fibrous filters. *Dokl. Akad. Nauk SSSR* **167**, 1327.

Stechkina, I. B. (1966). Diffusion precipitation of aerosols in fibrous filters. *Dokl. Akad. Nauk SSSR* **167**, 1327–1330 (in Russian).

Stechkina, I. B., and Fuchs, N. A. (1966). Studies on fibrous aerosol filters, 1: Calculations of diffusional deposition of aerosols in fibrous filters. *Ann. Occup. Hyg.* **9**, 59–64.

Strauss, W. (1959). Ph.D. thesis, Univ. of Sheffield.

Strauss, W. (1975). "Industrial Gas Cleaning," 2nd Ed. Pergamon, Oxford.

Subramanyam, W. V., and Kuloor, N. R. (1969). Dust deposition on single cylinders. *Ann. Occup. Hyg.* **12**, 9–25.

Thom, A. (1933). Flow past circular cylinders at low speeds. *Proc. R. Soc. London Ser A* **41**, 651–669.

Torgeson, W. L. (1958). General Mills, Rep. 1890.

Wong, J. B., and Johnstone, H. F. (1953). Univ. Illinois, Urbana, Eng. Exp. Sta. Rep. No. 11.

Wong, J. B., Ranz, W. E., and Johnstone, H. F. (1955). Intertial impaction of aerosol particles on cylinders. *J. Appl. Phys.* **26**, 244–249.

5

Gas Filtration

5.1 Introduction

Scope of Filters

Filters, along with wet scrubbers and electrostatic precipitators, are mostly used to remove particulates in the particle size range that is too small for cyclones to handle, that is, roughly the sizes under 10 μm. There are cases, however, where filters are used to remove particulates that are coarser than this value, given here only as an approximate size. The lower limit of particle sizes that can be removed by some highly efficient filters is not known with certainty. There are filters that are very efficient in removing particles on the order of 0.01 μm. Very comprehensive information on filtration is available in (Matteson and Orr, 1987).

Performance Criteria

There are three major performance criteria for a filter: collection efficiency, pressure loss, and lifetime. Collection efficiency tends to change with the filtering time because of the dust captured by the filter. Hence, it is customary to distinguish between instantaneous and cumulative collection efficiencies. Unlike the case of gravity settlers and cyclones, grade efficiency curves are not usually determined for filters. Usually, a filter is expected to have an overall collection efficiency of at least 99%. Normally, filters are operated at relatively low pressure drops, limited by the characteristics of the blower used to move the gas. The pressure loss increases with the amount of dust captured by the filter.

Types of Filter Media

Filter media for gas filtration mostly consist of woven fabrics, felt, loosely packed fibers, and paper, all of which consist of fibers. In some instances granular filters, wire screens, etc., are also used in gas filtration. Membrane or porous filters contain pores in a solid matrix and have special uses (e.g., Millipore filters). Nucleopore filters are very thin ($\sim 1.5\ \mu m$) sheets containing uniform, circular holes.

Classification of Filtration and Industrial Gas Filters

Filtration may be classified into two broad categories on the basis of where in the filter the filtering action is. In one category, most, if not all, of the filtering is done by a layer of the dust, the so-called filter cake, built up on the face of the filter. Before the build-up of the cake, these filters are rather ineffectual. In continuous operation, most of the deposited dust cake is periodically removed from the filter. In this kind of filtration, one attempts to prevent penetration of dust into the medium and resulting "blinding" of the filter. This kind of filtration is usually called *cake filtration*.

In the second category, the filtration takes place inside the filter medium. There is not supposed to be any buildup of a filter cake in this case. This mechanism of filtration is generally called *deep filtration*.

In industrial gas filtration there are three basic types of filters.

1. Woven or felted fabrics are in the form of bags, envelopes, or sleeves and are called *bag* or *fabric* filters. Conventional *bag houses* depend on the buildup of a dust cake for filtration. In some instances, however, bag filters also use depth filtration. In bag houses, most of the deposited dust is periodically removed from the bags, and filters are reused many times.
2. Fibrous filters are usually fabricated in the form of mats or screens composed of fibers, filaments, or ribbons, and have thicknesses ranging from 3 mm to 10 cm. They may also be fabricated in roll or pleated form. These are usually depth filters that are thrown away once they are plugged.
3. Bulk granular material is packed in a column, which is called an "aggregate" or "porous bed" filter. These are depth filters that often can be regenerated.

Before discussing the different kinds of filters, the relationship between pressure loss and the filtering velocity is examined, because it has a wide range of applicability to the various kinds of filters.

5.2 Pressure Loss in Filters

Darcy's Law

Filters belong to a broad class of materials called *porous media*. The theory of gas flow through porous media has been reviewed and discussed in many books (e.g., Carman, 1956; Collins, 1961; Dullien, 1979; Scheidegger, 1957; Greenkorn, 1983). At low (collector) Reynolds numbers, that is, in creeping or viscous flow, the relationship between flow rate and pressure drop, Δp, in porous media is given by Darcy's law:

$$v_s = \frac{K}{\mu}\frac{\Delta p}{L} \tag{5.1}$$

Here, L is the depth or thickness of the porous medium, sometimes referred to as "bed"; v_s is the superficial or face velocity, that is, the volume flow rate of the gas divided by the total normal cross-sectional area of the bed, including the solids; and K is the so-called Darcy permeability, which measures the inverse resistance, that is, the conductivity to flow, of the porous medium. Its dimension in the international system of units is m^2, and its practical unit is the darcy (1 darcy $= 9.87 \times 10^{-13}$ m^2).

The American Society for Testing and Materials (ASTM) definition of permeability for a cloth filter is the volume of air in cubic feet that will pass through 1 ft^2 (0.093 m^2) of a clean, new cloth in 1 min at a pressure drop of 0.5 in. (1.27 cm) water column. These permeabilities are different from the Darcy permeability, K, and are reported as pure numbers. There are numerous empirical and semiempirical equations in the literature for estimating the permeability of filter media.

Estimation of Permeability of Clean Filters

The most widely used relationship to estimate permeabilities was developed by Carman (1956) and is based on a flow model that visualizes channels or capillaries in a solid matrix:

$$K = \varepsilon^3/k_1 S^2(1-\varepsilon)^2 \tag{5.2}$$

Here, ε is the porosity or void fraction of the bed, that is, the volume of the free space contained in the bed divided by the total volume of the bed; S is the specific surface, based on volume of the solids in the bed; and k_1 is an empirical constant (Kozeny constant), which has been found to be equal to about 5 in many, but not all systems. The value of k_1 varies with the porosity very rapidly (Kyan *et al.*, 1970), as shown in Fig. 5.1, when the porosity exceeds 0.96.

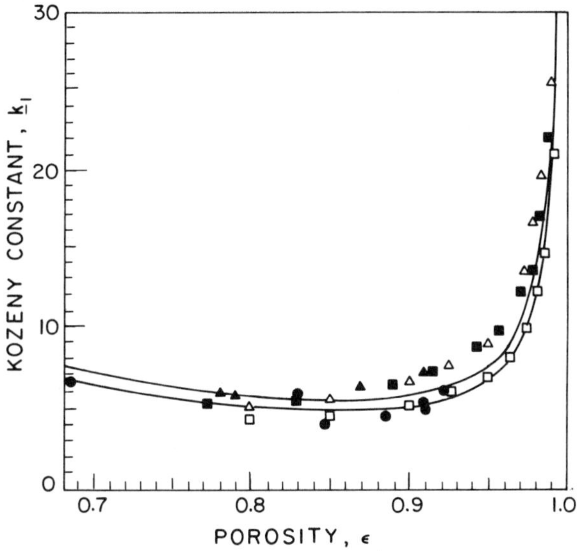

Figure 5.1 Variation of the Kozeny constant with porosity in fibrous beds (Kyan *et al.*, 1970).

A modification of Eq. (5.2) for porosities less than 0.88, which allows for the orientation of the fibers, is (Sullivan and Hertel, 1940)

$$K = \frac{k_3}{k_2} \frac{\varepsilon^3}{S^2(1-\varepsilon)^2} \tag{5.3}$$

Here, the ratio k_2/k_3 is 6.04 when fibers are normal to the direction of flow; 3.07 for fibers that are parallel to the direction of flow, and 5.5 for the usual case of a fiber blanket, where most of the fibers are normal to the direction of flow.

For beds of very fine fibers that are less than about 0.1 μm in diameter, Pich (1969) has suggested this expression for permeability:

$$K = D\lambda/4.58(1-\varepsilon) \tag{5.4}$$

where λ is the mean free path of the gas molecules. Equation (5.4) may be used only for $\varepsilon > 0.9$.

This empirical correlation has been suggested by Davies (1952):

$$K = \frac{D^2}{64(1-\varepsilon)^{3/2}\left[1 + 56(1-\varepsilon)^3\right]} \tag{5.5}$$

which can be used in the wide porosity range of about $0.4 < \varepsilon < 0.96$.

Davies has also suggested a correlation for fiber mats with porosities greater than 0.98, based on measurements of wool, cotton, rayon, glass, and steel wool pads, where the fiber size varied from 0.8 to 40 μm:

$$K = \frac{D^2}{70(1-\varepsilon)^{3/2}\left[1 + 52(1-\varepsilon)^{3/2}\right]} \tag{5.6}$$

The approach to the problem of estimating permeability, exemplified by Eqs. (5.2) and (5.3), may be referred to as *channel theory*, since the empty spaces are regarded as pores or channels in this approach. Another way of looking at the problem of flow through filters is to consider the resistance or drag offered to flow by the fibers immersed in the gas. This approach is usually referred to as *fiber drag theory*. Only the fiber drag theory may give the correct value of the permeability as the porosity approaches unity.

Among the equations using the fiber drag theory, only the one from Chen (1955) is mentioned here:

$$K = 0.129\frac{D_s^2\varepsilon}{(1-\varepsilon)}\ln\left[\frac{0.64}{(1-\varepsilon)^{1/2}}\right] \tag{5.7}$$

Equation (5.7) cannot be used for porosities under ~ 0.7. Here, D_s is the mean fiber diameter based on the surface area of the fibers.

In addition to the permeability, another useful quantity is the dimensionless coefficient of resistance, X, proposed by Davies (1952):

$$X = a^2\,\Delta p/\eta v_s L \tag{5.8}$$

which is related to the Darcy permeability as

$$X = a^2/K \tag{5.9}$$

(a = fiber radius). Various models of fibrous filters, predicting X, consist of arrays of infinitely long cylinders of certain specified orientations.

The pressure drop Δp has also been expressed as

$$\frac{\Delta p}{L} = F^* \frac{\mu v_s \beta}{\pi a^2} \tag{5.10}$$

[Iberall (1950), Happel and Brenner (1964), and Dawson (1969)], where $\beta = 1 - \varepsilon$ is the packing density of the filter, and F^* is a dimensionless drag defined by

$$F^* = \tfrac{1}{2} C_D \,\mathrm{Re}_c \tag{5.11}$$

which has also been expressed as

$$F^* = \frac{4\pi}{2 - \ln \mathrm{Re}_c} \tag{5.12}$$

Some of the theoretical equations for the filter resistivity X, as defined by Eq. (5.8) are given now. For transverse flow through a system of parallel cylinders,

$$X = \frac{4\beta}{-(1/2) - (1/2)\ln\beta + \beta^2/(1+\beta^2)} \tag{5.13}$$

[Langmuir (1942), Happel (1959)].

For cylinders equipartitioned in three mutually perpendicular directions, one of which is in the direction of flow,

$$X = \frac{4\beta}{3(1-\beta)} \frac{4 - \ln \mathrm{Re}_c}{2 - \ln \mathrm{Re}_c} \tag{5.14}$$

[Iberall (1950)].

For high porosity fibrous systems ($\beta \ll 1$),

$$X = \frac{4\beta}{-0.25 - 0.5\ln\beta} \tag{5.15}$$

[Fuchs and Stechkina (1963), Billings (1966)].

EXAMPLE 5.1

Compare the various predicting equations of permeability K. Assume cylindrical fibers of uniform diameter $D = 1\ \mu$m and vary the void fraction ε.

SOLUTION

The specific surface, S, of a cylinder is $S = 4/D$. Thus, Eq. (5.2) takes the form

$$K = \frac{\varepsilon^3 D^2}{80(1-\varepsilon)^2}$$

and Eq. (5.3) takes the form

$$K = \frac{k_3}{k_2} \frac{\varepsilon^3 D^2}{16(1-\varepsilon)^2}$$

Equations (5.4)–(5.7) are tailor-made for cylindrical collectors.

	$K \times 10^{15}$ (m^2)						
ε	0.5	0.6	0.7	0.8	0.9	0.95	0.99
Eq. (5.2)	✓6.25	✓17.0	✓48	✓160	✓910	4,300	121,000
Eq. (5.5)	✓5.50	✓13.0	✓38	✓120	✓470	✓1,400	✓16,000
Eq. (5.6)	2.10	4.0	9	28	170	810	✓14,000
Eq. (5.7)	−13	2.0	✓47	✓180	✓820	✓2,570	✓24,000

The values with the check marks are reasonably consistent and, therefore, are probably of the correct order of magnitude. Negative values are without physical meaning.

5.3 Bag (Fabric) Filters

Bag filters are extensively employed to control harmful or obnoxious emissions in operations involving *abrasives*; *irritating chemical dusts*; and the *exhausts from electric furnaces*, *oil fired boilers*, and *oxygen fed converters for steel making*. The basic filter element consists of a woven or felted textile material in the shape of a tube or sleeve.

The principle of operation of a conventional bag filter consists mainly of the filtering effect of a particle layer (dust cake) that acculates on the fabric. At the same time, when the dust cake starts building up, the fabric must not "blind," that is, the void space between the fibers must not be packed full with dust, because this would result in a greatly diminished permeability of the filter. This, however, is possible only if the surface of the clean cloth already captures dust particles with a relatively high efficiency, resulting in a rapid reduction of the surface void space of the fabric, while in the interior, the void space is affected to a much lesser degree. The larger particles present in the dust get captured preferentially by the surface of the clean cloth and are, therefore, chiefly responsible for the start of the dust cake buildup. The finer dust particles are captured mostly by the dust cake and thus blinding of the fabric is avoided. The filtering action of the dust layer has not been successfully analyzed theoretically. When the pressure loss increases because of the accumulated load, and simultaneously the gas velocity decreases, the filtering operation must be stopped at a prescribed point, and the filter cleaned by dust dislodging operations; then the filter is reused over and over again. Bag filters are generally used with gas streams of dust loadings of 0.1–100 g m^{-3}. Many bag filters are assembled in a single chamber or compartment that is commonly known as a *bag house*. Each bag house has its own gas inlet and outlet, dampers or valves, hopper, and cleaning system.

In Table 5.1, the basic properties of various fibers, which are presently widely used, are listed; and in Fig. 5.2, several weave types are shown. The performance of any fabric is greatly influenced by thread density, fiber composition, and nap, which is the hairy or downy texture of the cloth

Figure 5.2 Filter cloth weaves (Iinoya and Orr, 1977).

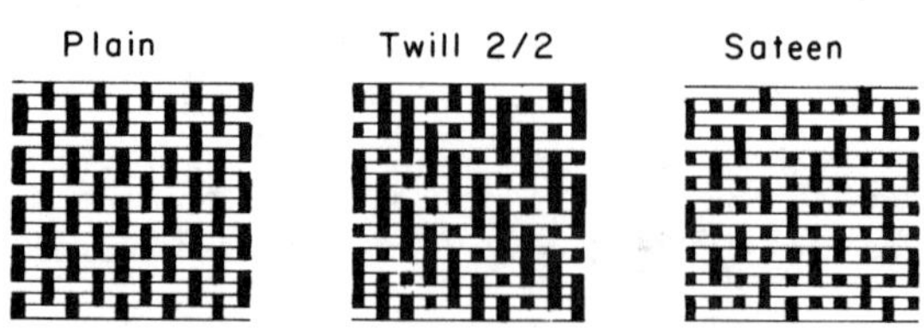

Table 5.1 Properties of Fiber Materials[a]

Fiber	Physical characteristics				Relative resistance to attack by			Other attribute
	Relative strength	Specific gravity	Normal moisture content (%)	Maximum usable temperature (°F)	Acid	Base	Organic solvent	
Cotton	Strong	1.6	7	180	Poor	Medium	Good	Low cost
Wool	Medium	1.3	15	210	Medium	Poor	Good	
Paper	Weak	1.5	10	180	Poor	Medium	Good	Low cost
Polyamide (nylon)	Strong	1.1	5	220	Medium	Good	Good[b]	Easy to clean
Polyester (Dacron)	Strong	1.4	0.4	280	Good	Medium	Good[c]	
Acrylonitrile (Orlon)	Medium	1.2	1	250	Good	Medium	Good[d]	
Vinylidene chloride	Medium	1.7	10	210	Good	Medium	Good	
Polyethylene	Strong	1.0	0	250	Medium	Medium	Medium	
Tetrafluoroethylene	Medium	2.3	0	500	Good	Good	Good	Expensive
Polyvinyl acetate	Strong	1.3	5	250	Medium	Good	Poor	
Glass	Strong	2.5	0	550	Medium	Medium	Good	Poor resistance to abrasion
Graphitized fiber	Weak	2.0	10	500	Medium	Good	Good	Expensive
Asbestos	Weak	3.0	1	500	Medium[e]	Medium	Good	
"Nomex" nylon	Strong	1.4	5	450	Good	Medium	Good	Poor resistance to moisture

[a]From Iinoya and Orr (1977).
[b]Except phenol and formic acid.
[c]Except phenol.
[d]Except heat.
[e]Except SO_2.

surface. Cotton is the least expensive fiber. Silicone- or graphite-coated glass cloth is commonly employed up to 250°C.

The electrostatic properties of both the dust and the collecting fabric may interfere with the filtering and cleaning performance. Antistatic fabrics have been developed to minimize the effects of changes.

The superficial or face velocity in bag houses is commonly referred to as the gas-to-cloth ratio, that is, the m^3/min, or ft^3/min, of gas filtered per m^2 or ft^2 of filter area, and is often expressed as mm/s or ft/min. The choice of the gas-to-cloth ratio in practical operation depends on a number of factors. Higher velocities give greater penetration of the cloth and make cake removal more difficult; they also tend to increase the pressure drop across the filter. Excessively low filtration rates, however, increase the filter area required, and consequently, bigger plants are required to handle the same volume. The gas-to-cloth ratios used vary from 5 to 75 mm s^{-1}. Typical ratios for a number of materials are listed in Table 5.2.

Types of Bag Filters

As can be seen in Table 5.2, there are various types of bag filters, such as shakers, reverse air collectors, and pulse-jet collectors. There are two broad categories of shaker collectors: those for intermittent use and those for continuous use. The difference is that continuous use collectors are compartmentalized to allow a portion of the whole collector to be taken off stream for cleaning, while dirty gas continues to be filtered by the rest of the unit. Intermittent use collectors usually have a single compartment and are smaller in size than the continuous use collectors.

It has been found that it is necessary to stop completely the forward flow of gas through the cloth in order to break up the dust cake and keep the dislodged dust from reclinging to or reentraining on the cloth. Hence, a single compartment unit requires periodic shut down for shaking, even though this process may require only a few minutes.

Because shakers have been in existence since before the turn of the century, a great deal of research has been done on this type of collector. The main parameters in shaker mechanism design are frequency (number of shakes per minute), amplitude (distance over which the top of the tube is moved), acceleration (to which the cloth is subjected), and duration of shaking. A rectangular, three-compartment bag house with a circular filter sleeve is shown in Fig. 5.3, and a diagram showing the various methods of shaking cylindrical filter sleeves is shown in Fig. 5.4. Shakers are commonly available in the largest range of size of any type of fabric filter. There are small, manually operated units for intermittent duty, and there are also

Table 5.2 Typical Air-to-Cloth Ratios[a]

	Usual air–cloth ratio (cfm ft^{-2})[b]		
Dust	Shaker collector	Pulse-jet collector	Reverse air collector
Alumina	2.5–3.0	8–10	—
Asbestos	3.0–3.5	10–12	—
Bauxite	2.5–3.2	8–10	—
Carbon, black	1.5–2.0	5–6	1.1–1.5
Coal	2.5–3.0	8–10	—
Coal fly ash[c]	—	4–6	1.5–2.5
Cocoa, chocolate	2.8–3.2	12–15	—
Clay	2.5–3.2	9–10	1.5–2.0
Cement	2.0–3.0	8–10	1.2–1.5
Cosmetics	1.5–2.0	10–12	—
Enamel frit	2.5–3.0	9–10	1.5–2.0
Feeds, grain	3.5–5.0	14–15	—
Feldspar	2.2–2.8	9–10	—
Fertilizer	3.0–3.5	8–9	1.8–2.0
Flour	3.0–3.5	12–15	—
Graphite	2.0–2.5	5–6	1.5–2.0
Gypsum	2.0–2.5	10–12	1.8–2.0
Iron ore	3.0–3.5	11–12	—
Iron oxide	2.5–3.0	7–8	1.5–2.0
Iron sulfate	2.0–2.5	6–8	1.5–2.0
Lead oxide	2.0–2.5	6–8	1.5–1.8
Leather dust	3.5–4.0	12–15	—
Lime	2.5–3.0	10–12	1.6–2.0
Limestone	2.7–3.3	8–10	—
Mica	2.7–3.3	9–11	1.8–2.0
Paint pigments	2.5–3.0	7–8	2.0–2.2
Paper	3.5–4.0	10–12	—
Plastics	2.5–3.0	7–9	—
Quartz	2.8–3.2	9–11	—
Rock dust	3.0–3.5	9–10	—
Sand	2.5–3.0	10–12	—
Sawdust (wood)	3.5–4.0	12–15	—
Silica	2.3–2.8	7–9	1.2–1.5
Slate	3.5–4.0	12–14	—
Soap, detergents	2.0–2.5	5–6	1.2–1.5
Spices	2.7–3.3	10–12	—
Starch	3.0–3.5	8–9	—
Sugar	2.0–2.5	7–10	—
Talc	2.5–3.0	10–12	—
Tobacco	3.5–4.0	13–15	—
Zinc oxide	2.0–2.5	5–6	1.5–1.8

[a] From Kennedy and Cheremisinoff (1977).

[b] Values tabulated are based on light to moderate loadings (~ 2 g m^{-3}) of dust having particle size and shape characteristics typical of the specific material. Ratios will normally be less when dust loading is very heavy, temperature is elevated, or particle size is smaller than commonly encountered (adapted from American Air Filter Co. Data Sheet).

[c] Glass fabrics frequently used.

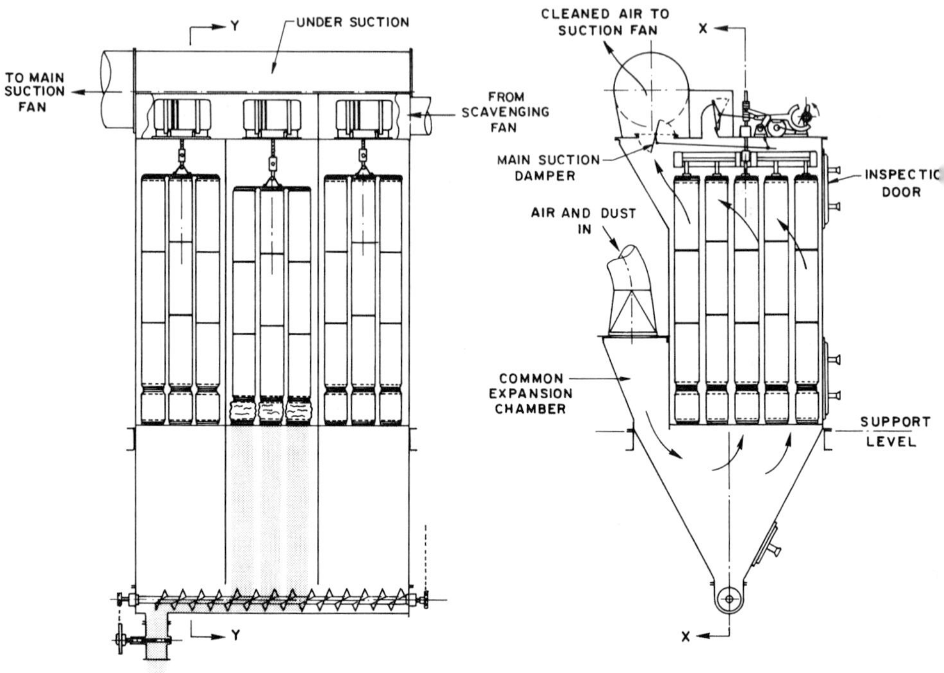

Figure 5.3 Rectangular, three-compartment bag house with circular filter sleeves. (Courtesy of Visco Engineering Co. Ltd., Croydon, Surrey.)

systems designed for continuous service and the handling of millions of cubic feet per minute.

Reverse air collectors are those which clean the filter medium by stopping the forward flow of gas and introducing a low-pressure flow in the opposite direction. Differential pressure accomplishes the cleaning by collapsing the filter medium and breaking up the dust cake.

The most common fabric configuration for this type of unit is the tube. On large units, these tubes may be as much as one foot in diameter and up to sixty feet in length. Woven glass is used very often, since it provides the ideal property of high tensile strength at temperatures to 400°C.

Reverse air collectors are usually compartmentalized to permit one section at a time to be taken off stream for cleaning. Cleaning durations must be long enough to permit particles from the topmost part of the tube to fall into the hopper.

Filtration velocities of these types of collectors are usually very low. More cloth is needed for a given flow, and hence, reverse air collectors are substantially larger than other types.

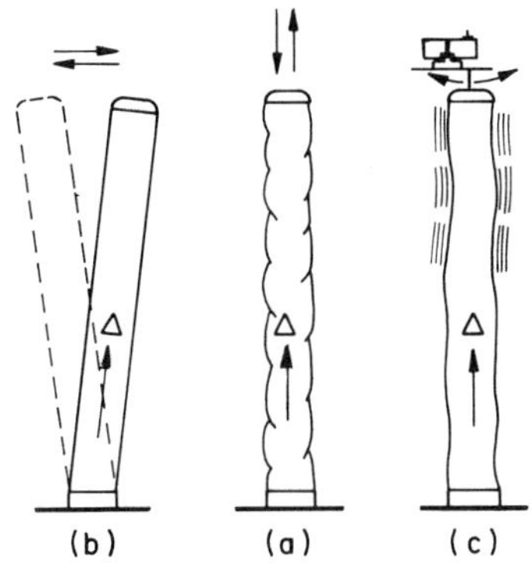

Figure 5.4 Methods of shaking cylindrical filter sleeves: (a) vertical filter sleeve movement, (b) gentle sideways movement, (c) sideways vibration of filter sleeve. [Reprinted with permission from Strauss (1975).]

Pulse-jet collectors contain a number of closed-end filter tubes, supported at the top and held open by wire cages. Dust laden air is passed through the tubes, and the dust is deposited on the outside of the cloth. The tubes are cleaned by a brief burst of compressed air injected into the tube at the top through a venturi. The main advantage offered by this design is that it allows a significant increase in air-to-cloth ratio. The pulse-jet collector enjoys the status of the ultimate form of fabric filter.

Operating Conditions of Bag Houses

When a new filter cloth is first put into operation, it does not contain any dust, but after dirty gases have passed through the cloth and it has been cleaned several times, an approximately constant amount of collected dust remains in the filter. This is called the *equilibrium dust content* of the cloth, x_e. The amount of this depends on the type of filter material, the dust size, and the timing and type of the filter cake removal system. An operating pressure drop of 0.75–1.5 kPa is standard for most bag houses, but some bag houses can run at substantially higher or lower drops.

Darcy's law is usually written in this form for fabric filters:

$$\Delta p = v_s(a + bx) \qquad (x \geq x_e) \tag{5.16}$$

where $(a + bx)$ is the resistance of the dusty filter cloth to the flow of gas of viscosity μ [i.e., $(\mu L/K)$ in Eq. (5.1)], a and b are empirical constants, and x is the total dust content of the cloth, measured in g m^{-2} of cloth. Equation (5.16) expresses the fact that the pressure drop Δp is proportional to the superficial velocity v_s and the resistance $(a + bx)$. It is valid only for a narrow range of values of x, and it is not valid below $x = x_e$. Hence a is not equal to the resistance of the clean cloth. The dust content x of the cloth can be expressed as

$$x = x_e + L_i V \tag{5.17}$$

where x_e (g m^{-2}) is the equilibrium dust content of the cloth, L_i (g m^{-3}) is the dust loading, and V (m^3 m^{-2} cloth) is the volume of as filtered after the last cleaning of the filter.

The amount of gas filtered as a function of time is readily calculated under conditions of constant velocity. In a constant pressure operation one lets

$$v_s = dV/dt \tag{5.18}$$

and, after combining Eqs. (5.18) and (5.17) with Eq. (5.16),

$$\frac{dV}{dt} = \frac{\Delta p}{a + b(x_e + L_i V)} \tag{5.19}$$

Integrating Eq. (5.19) at constant Δp, between the limits $t = 0$ and t and $V = 0$ and V, gives this relationship between time and the volume of gas filtered:

$$t = (V/\Delta p)(a + bx_e + bL_i V/2) \tag{5.20}$$

In an actual example, the data found were

$$a = -19.6 \text{ kPa s m}^{-1}$$

$$b = 0.137 \text{ kPa s m g}^{-1}$$

$$x_e = 172 \text{ g m}^{-2}$$

$$L_i = 4.58 \text{ g m}^{-3}$$

Table 5.3 Instantaneous and Average Filtering Velocities for a Typical Bag Filter Installation Operating at Constant Pressure Drop[a]

Volume of gas filtered per m^2 of cloth (m^3)	Time required (min)	Face velocity (mm s^{-1}) at time t (min)	Average face velocity (mm s^{-1}) after time t (min)
0	0	64	64
3.05	1	43	51
6.10	2.4	32	43
9.15	4.1	26	37
10.67	5.2	23	34
15.25	8.9	18	28
18.3	11.8	16	26
30.5	27.5	11	18

[a] Reprinted with permission from Strauss (1975).

and

$$\Delta p = 0.249 \text{ kPa}$$

The results of the calculations, using Eq. (5.20), are shown in Table 5.3, which shows that the face velocity v_s drops rapidly as the resistance of the cloth builds up.

In a constant rate operation, it is the pressure buildup as a function of time that is of interest. The calculation of the pressure buildup is shown in Example 5.2.

EXAMPLE 5.2: CONSTANT RATE FILTRATION

Consider a fabric (bag) filter installation which is to be used to filter gas of a dust burden of 5 g m^{-3}. The following relation has been found to apply empirically for this system between the values of the dust content $x_e = 200$ g m^{-2} and $x = 400$ g m^{-2}.

$$\Delta p \text{ (kPa)} = v_s \text{ (m s}^{-1}) \left[-15 + 0.1x \text{ (g m}^{-2})\right]$$

where

$$x = x_e + L_i V$$

It is planned to run this filter at a constant, fixed air-to-cloth ratio. Calculate the pressure drop and the time elapsed after filtering through the

Figure 5.5 Velocity pattern in a six-compartment bag house as a function of time. Average velocity, 1 m min^{-1}; initial compartment pressure loss, 17 mm H_2O; final compartment pressure loss, 170 mm H_2O (Iinoya and Orr, 1977).

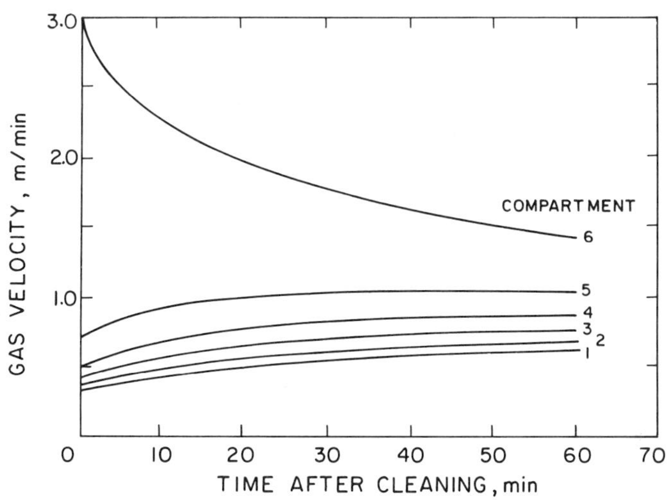

freshly cleaned filter ($x = x_e$) 10, 20, 30, and 40 m^3 of air per square meter of cloth, at a constant air-to-cloth ratio of 0.05 m^3 s^{-1}/m^2.

SOLUTION

We have the formulas for bag filters [Eqs. (5.8) and (5.9)]:

$$\Delta p = v_s(a + bx)$$

$$x = x_e + L_i V$$

At constant v_s,

$$V = v_s t \qquad \text{and} \qquad x = x_e + L_i v_s t$$

$$\Delta p = v_s[a + b(x_e + L_i v_s t)] = v_s(a + bx_e) + L_i b v_s^2 t$$

$$\therefore \Delta p = 0.05(-15 + 0.1 \times 200) + 5 \times 0.1 \times 25 \times 10^{-4} t$$

$$\therefore \Delta p = 0.25 + 1.25 \times 10^{-3} t$$

The results are tabulated here

$V\left(\frac{m^3}{m^2}\right)$	t (s)	Δp (kPa)
10	200	0.50
20	400	0.75
30	600	1.00
40	800	1.25

In the actual operation of multicompartment bag houses, usually neither the pressure drop nor the gas flow rate is constant. The typical pressure

Figure 5.6 Flow and pressure loss variations as a function of time in a multicompartment bag house. Average velocity, 1 m min^{-1}; initial compartment pressure loss, 17 mm H_2O; final compartment pressure loss, 170 mm H_2O (Iinoya and Orr, 1977).

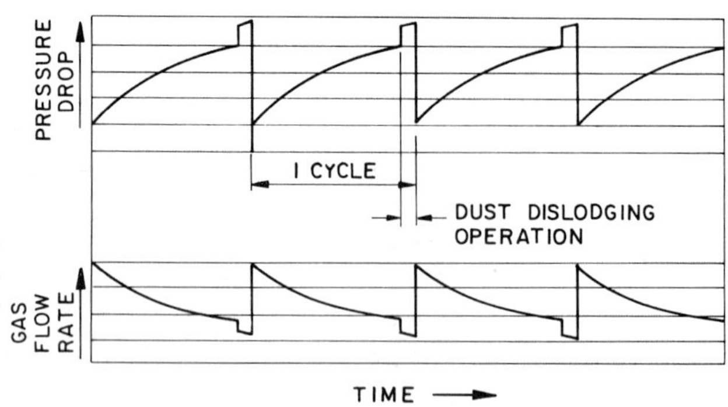

Table 5.4 Relative Fabric Cost Factors

Fiber type	Relative cost
Cotton	1.0
Polypropylene	1.75
Polyester (Dacron)	2.8
Acrylic (Dynel)	3.2
Glass	5.5
Polyamide (Nomex)	8.0
Poly TFF (Teflon)	
Felted	30–71
Woven	21

Figure 5.7 Purchase cost of fabric filters (1970). A, High-temperature synthetic, woven and felt; continuously automatically cleaned. B, Medium-temperature synthetic, woven and felt; continuously automatically cleaned. C, Woven natural fibers; intermittently cleaned, single compartments. Costs for equipment of indicated construction may vary by 20% of reported figure (Edmisten and Bunyard, 1970).

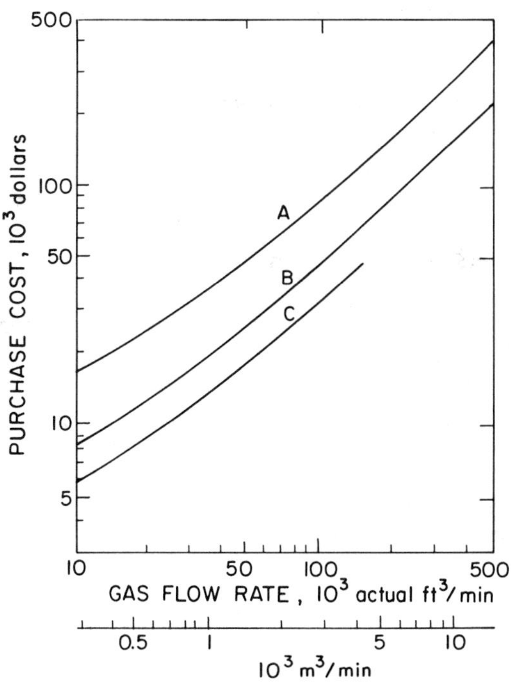

drop and gas flow rate behavior in a multicompartment bag house is shown schematically in Figs. 5.5 and 5.6.

Bag houses provide about 99% efficiency for removal of particles larger than 1 μm, but the failure of even a single filter bag may seriously reduce the overall efficiency. The use of fabric filters constitutes one of the more costly methods of dust collection, but this is offset by high efficiency and the ability of fabric filtration to meet the most stringent particulate emission codes. Relative costs of various fabrics are shown in Table 5.4 (Iinoya and Dennis, 1987). Figure 5.7 shows a typical relationship between purchase cost and gas flow rate. Maintenance costs are applicable because filter media must be replaced regularly during the life of the system.

5.4 Fibrous Filters

Operating Conditions of Fibrous Filters

Fibrous filters are used for a large variety of different purposes, for example, to prefilter gas streams to remove a significant fraction of the dust burden. They may be fabricated in panel or continuous roll forms. Generally, the objective is to remove the coarser particles at relatively low pressure losses. Paper filters are commonly used to provide either high-efficiency precleaning, where 90–95% of the atmospheric dust is captured, or ultra high-efficiency collection, where ~ 99.95% removal of 0.25 μm particles may be attained. For high temperatures, the medium may be made of glass or mineral fibers. Fibrous filters are used extensively in air conditioning, heating, and ventilating systems, but high-efficiency fibrous filters are also employed as after cleaners. For fibrous mat filters, the superficial velocities range from very low values to values that are much higher than those for bag filters: 1 m s^{-1} may be regarded as a representative value for a high-velocity application.

An empirical relationship has been proposed for the pressure loss of clean fibrous mats (Iinoya and Orr, 1977):

$$\Delta p = \left(0.6 + \frac{4.7}{\sqrt{\mathrm{Re_c}}} + \frac{11}{\mathrm{Re_c}}\right)\frac{2\rho v^2 L(1-\varepsilon)}{\pi D} \tag{5.21}$$

This equation is generally satisfactory for Reynolds numbers from 10^{-3} to 100 and void fractions $\varepsilon = 0.92$–0.98. Here, the interstitial gas velocity v is

$$v = v_s/\varepsilon \tag{5.22}$$

The filtering mechanism of fibrous filters is reasonably well understood; nevertheless, it is not possible to accurately design such filters from scratch using single fiber efficiencies, because of numerous complicating factors. In media filtration no dust cake is formed. The clean filter must already be fully efficient in this case.

The use of fibrous filters for the removal of fine, uncharged particulate matter suspended in a gas stream is mainly due to inertial effects, interception, and Brownian diffusion. It should be noted that not all the particulates captured by a fiber adhere to it and are, therefore, removed from the gas stream. A considerable fraction bounce back from the fiber into the gas stream, and as a result, the theoretical single fiber collection efficiencies often tend to be too high.

Collection Efficiency of a Fibrous Filter

The theory discussed here has been proposed for calculating the collection efficiency of a fiber mat. Even though this theory cannot be used to predict filter efficiencies quantitatively, it does provide a good insight into the mechanism of filtration, and therefore, it is a highly valuable aid in understanding deep filtration.

Consider a very thin section of thickness dL of a fibrous mat of total cross-section A. The plane of the section is perpendicular to the direction of flow; dL should be much greater than the fiber diameter, but small enough to approximate a single layer of fiber in the plane of the section (i.e., negligible extent of overlap of fibers). Let L_{f} be the length of the fiber per unit volume of the mat. Then the length of the fiber in the thin section is $L_{\mathrm{f}} A\, dL$.

The projected area of the fibers on the plane of the section is $(\pi/4) D L_{\mathrm{f}} A\, dL$, where D is the fiber diameter. Random orientation of the fibers in the mat is assumed.

Using the definition of total (including all mechanisms of collection) collection efficiency, $\eta_{\mathrm{f}\beta}$, of a fiber embedded in a filter, one can write

$$\eta_{\mathrm{f}\beta} = \frac{\text{number of particles deposited in thin section in unit time}}{N v (\pi/4) D L_{\mathrm{f}} A\, dL} \qquad (5.23)$$

where v is the interstitial velocity [see Eq. (5.22)] and N is the number of particles per unit volume of gas. The denominator in Eq. (5.23) gives the number of particles crossing the area $(\pi/4) D L_{\mathrm{f}} A\, dL$ in unit time.

The number of particles deposited in the thin section in unit time, dn, can be expressed by mass balance, independently of Eq. (5.23), as

$$dn = -dN\, v_{\mathrm{s}} A \qquad (5.24)$$

where $-dN$ is the change in particle concentration of the gas between entering and leaving the thin section and $v_{\mathrm{s}} A$ is the volume flow rate of the gas. Upon introducing Eq. (5.24) into Eq. (5.23), after simplification, using Eq. (5.22) and rearrangement, one obtains

$$-\frac{dN}{N} = \frac{\pi}{4}\frac{\eta_{f\beta}DL_f}{(1-\beta)}\,dL \tag{5.25}$$

which gives, on integration between the limits N_{in} and N_{out}, 0 and L, where L is the depth of the mat, the expression

$$\ln\frac{N_{out}}{N_{in}} = -\frac{\pi}{4}\frac{\eta_{f\beta}DL_fL}{(1-\beta)} \tag{5.26}$$

It is convenient to eliminate L_f from Eq. (5.26) by using the identity

$$\beta = \pi D^2L_f/4 \tag{5.27}$$

This results in the final expression:

$$\ln\frac{N_{out}}{N_{in}} = -\frac{\eta_{f\beta}L\beta}{D(1-\beta)} \tag{5.28}$$

Equation (5.28) states that the particle concentration in the gas stream, while passing through the fibrous mat, decreases exponentially with increasing bed depth L and individual fiber efficiency $\eta_{f\beta}$, and with decreasing fiber diameter D, provided that β is kept constant.

Equation (5.28) may be used directly to estimate the overall efficiency η_0 of a fibrous mat because of this relationship between η_0 and N_{out}/N_{in}:

$$\eta_0 = \frac{N_{in}-N_{out}}{N_{in}} = 1-\frac{N_{out}}{N_{in}} \tag{5.29}$$

Hence, by Eqs. (5.28) and (5.29), we obtain for η_0 the formula

$$\eta_0 = 1-\exp\left[-\frac{\eta_{f\beta}L\beta}{D(1-\beta)}\right] = 1-e^{-\lambda L} \tag{5.30}$$

where $\lambda = \eta_{f\beta}\beta/D(1-\beta)$ is the filtration coefficient.

Methods of Estimation of Collection Efficiency of a Fiber, $\eta_{f\beta}$, in a Fibrous Filter

The collection efficiency of a fiber embedded in a filter, $\eta_{f\beta}$, differs from the collection efficiency of isolated single fibers, because the velocity fields around the fibers interfere with each other.

Davies (1952) concluded that the interference effect is the same for different filtration mechanisms and could be expressed for any filtration mechanism as

$$\eta_{f\beta} = \eta(0.16 + 10.9\beta - 17\beta^2) \tag{5.31}$$

where η is calculated by any one of Eqs. (4.15)–(4.17), (4.20), (4.22)–(4.24), and (4.39)–(4.59).

Chen (1955) concluded that the interference effect was positive and expressed the interference effect by the equation

$$\eta_{f\beta} = \eta(1 + 4.5\beta) \tag{5.32}$$

However, Dorman (1964) found these empirical relations:

$$\eta_{\mathcal{D}\beta} = \eta_{\mathcal{D}}(1 - 4\beta) \tag{5.33}$$

$$\eta_{C\beta} = \eta_C(1 + 30\beta) \tag{5.34}$$

and

$$\eta_{I\beta} = \eta_I(1 + 110\beta) \tag{5.35}$$

5.5 Aggregate and Porous Bed Filters

Aggregate filters may consist of beds of coke used for acid mist recovery, of gravel used for exhaust gas cooling as well as particle collection, or of ceramic, plastic, or metal shapes used for special applications. Countercurrent water spray, often containing additives, may be used for regenerating the filter.

Porous bed filters usually consist of sintered or compressed metal or ceramic granules, and are commonly used to clean compressed or high-temperature process gas streams for product or catalyst recovery.

EXAMPLE 5.3

A fibrous mat filter is made of fibers 10 μm in diameter and is to be used to filter 0.5-μm aerosol particles (density, 2.8 g cm^{-3}) from air at ambient conditions (assume spherical particles). A face velocity of 1 m s^{-1} is contemplated.

1. Estimate the individual fiber collection efficiency by all the mechanisms known to you; then form the combined efficiency of an individual fiber.

2. Estimate the permeability of the fibrous mat before use in μm^2 units if the void fraction is 0.8.
3. What is the efficiency of a fibrous mat of $\varepsilon = 0.8$ and $L = 2$ mm? Will this filter have a long life?
4. A fibrous mat filter of 99% efficiency is to be designed from the fibers. The thickness L and the void fraction ε are the design parameters. What is the void fraction for $L = 40$ mm? What is ε for $L = 100$ mm? Which of these two designs will have a longer life (i.e., will clog later)? Why?
5. A fibrous mat filter of 99% efficiency (clean filter) is to be designed, using the given fibers, for theoretically longest life, which is defined (in a first approximation) by the condition that every cross section of the clean filter perpendicular to the direction of gas flow collects the same amount of dust.
   a. Derive a general relationship for the filtration coefficient λ as a function of fractional depth $\xi = x/L$ of the filter for longest filter life. Calculate and tabulate λ versus ξ for $L = 30$ mm.
   b. Design a filter consisting of three layers of fibrous mat, each layer of a different void fraction, that would approximate the conditions of the theoretically longest filter life calculated under a. State the fractional thickness $\Delta\xi$ of each layer, the corresponding void fraction ε, and the fraction of dust collected by each layer.

SOLUTION

1. Here, μ_{IC} is calculated on the assumption that $v = v_s = v_\infty$.

$$\mathrm{Re_c} = \frac{v_\infty \rho D}{\mu} = \frac{10^{-3}\ \mathrm{cm} \times 100\ \mathrm{cm\ s^{-1}} \times 1.205 \times 10^{-3}\ \mathrm{g\ cm^{-3}}}{1.816 \times 10^{-4}\ \mathrm{P}}$$

$$\mathrm{Re_c} = 0.66 \qquad \text{(close to viscous flow)}$$

$$\Psi = \frac{C\rho_p d^2 v_\infty}{18\mu D}$$

$$\lambda = 6.56 \times 10^{-8}\ \mathrm{m}$$

$$C = 1 + \left(\frac{2 \times 6.56 \times 10^{-8}}{0.5 \times 10^{-6}}\right)\left[1.246 + 0.42 \exp\left(\frac{-0.8 \times 0.5 \times 10^{-6}}{2 \times 6.5 \times 10^{-8}}\right)\right]$$

$$\therefore C = 1.33$$

$$\therefore \Psi = \frac{(1.33)(0.5 \times 10^{-4})^2(100)(2.8)}{(18)(1.816 \times 10^{-4})(10 \times 10^{-4})} = 0.28$$

Assuming that Davies' formula [Eq. 4.24(a)] also applies at $\mathrm{Re_c} = 0.66$,

$$R = d/D = 0.5/10 = 5 \times 10^{-2}$$

$$\eta_{\mathrm{IC}} = 0.16[5 \times 10^{-2} + (0.5 + 0.5 \times 5 \times 10^{-2})0.28$$
$$- 0.1052 \times 5 \times 10^{-2} \times 0.28^2] = 3.2 \times 10^{-2} = 3.2\%$$

$$\mathscr{D} = \frac{kTC}{3\pi\mu d} = \frac{1.38 \times 10^{-23} \times 293 \times 1.33}{3\pi(1.816 \times 10^{-5})(0.5 \times 10^{-6})} = 6.38 \times 10^{-11}\ \frac{\mathrm{m}^2}{\mathrm{s}}$$

$$\mathrm{Pe} = \frac{v_\infty D}{\mathscr{D}} = \frac{1 \times 10 \times 10^{-6}}{6.38 \times 10^{-11}} = 1.57 \times 10^5$$

$$\eta_{\mathscr{D}} = \frac{1}{\mathrm{Pe}} + 1.727\frac{\mathrm{Re_c^{1/6}}}{\mathrm{Pe}^{2/3}} = 5.5 \times 10^{-4} = 0.05\% \qquad \text{[Eq. (4.43)]}$$

(negligibly small).

2. $$K = \frac{10^2\ \mu\mathrm{m}^2}{(64)(0.2)^{3/2}(1 + 56 \times 0.2^3)} = 12.06\ \mu\mathrm{m}^2 \qquad \text{[Eq. (5.5)]}$$

3. $$\eta_0 = 1 - \exp\left[-\frac{0.032 \times 2 \times 10^{-3} \times 0.2)}{10 \times 10^{-6} \times 0.8}\right] = 0.810 \qquad \text{[Eq. (5.30)]}$$

The filter will not have a long life, because it will clog early, owing to its small void space.

4. $$0.01 = \exp\left[-\frac{0.032(1-\varepsilon)}{10^{-2}\varepsilon}L\right]$$

Therefore,

$$\frac{(1-\varepsilon)}{\varepsilon}L = 1.44$$

$$L = 100\ \mathrm{mm} \qquad \varepsilon = 0.99$$
$$L = 40\ \mathrm{mm} \qquad \varepsilon = 0.97$$

The 100-mm-thick filter of $\varepsilon = 0.99$ will clog later because a layer of equal thickness will collect the smaller amount of dust in this filter.

5. Let us divide the filter into n parallel layers of equal thickness, $1, 2, 3, \ldots, (n-2), (n-1), n$, where the first layer is at the inlet and the nth layer is at the outlet face of the fibrous mat. Each layer will be characterized by a penetration,

$$\frac{N_2}{N_1}, \frac{N_3}{N_2}, \frac{N_4}{N_3}, \ldots, \frac{N_{n-1}}{N_{n-2}}, \frac{N_n}{N_{n-1}}, \frac{N_{n+1}}{N_n}$$

a. Applying the condition of longest life to the last layer,

$$N_n - N_{n+1} = (1/n)(N_1 - N_{n+1})$$

Therefore,

$$N_n = \frac{N_1 + (n-1)N_{n+1}}{n}$$

and

$$\frac{N_{n+1}}{N_n} = \frac{n(N_{n+1}/N_1)}{(n-1)(N_{n+1}/N_1) + 1}$$

Applying the same condition to the second to the last layer,

$$N_{n-1} - N_n = \frac{1}{n}(N_1 - N_{n+1})$$

Therefore,

$$N_{n-1} = \frac{2N_1 + (n-2)N_{n+1}}{n}$$

giving

$$\frac{N_n}{N_{n-1}} = \frac{(n-1)(N_{n+1}/N_1) + 1}{(n-2)(N_{n+1}/N_1) + 2}$$

Generalization gives

$$N_{n-j} = \frac{(j+1)N_1 + (n-1-j)N_{n+1}}{n}$$

and

$$\frac{N_{n-(j-1)}}{N_{n-j}} = \frac{(n-j)(N_{n+1}/N_1) + j}{(n-j-1)(N_{n+1}/N_1) + (j+1)}$$

The dust removal efficiency of the $(n-j)$th layer is

$$\eta_{n-j} = 1 - \frac{N_{n-(j-1)}}{N_{n-j}} = 1 - e^{-\lambda_{n,j}L/n}$$

Therefore,

$$\lambda_{n,j}L = -n \ln \frac{(n-j)(N_{n+1}/N_1) + j}{(n-j-1)(N_{n+1}/N_1) + (j+1)}$$

We seek the limit of this expression as $n \to \infty$, because this corresponds to continuous variation of λ from one face of the filter to the other. We write the previous expression as

$$\lambda_\xi L = \lim_{n \to \infty} \left\{ -n \ln \frac{(1-\xi)Pt + \xi}{[1 - \xi - (1/n)]Pt + [\xi + (1/n)]} \right\}$$

which is indeterminate ($\xi = j/n$). Using l'Hôpital's Rule, however, yields this result for the limit:

$$\lambda_\xi L = \frac{(1 - Pt)}{\xi + (1 - \xi)Pt} = \frac{\eta_0}{1 + \eta_0(\xi - 1)}$$

λ_ξ (mm^{-1})	ξ	
3.30	0	
1.66	0.01	B: 2% of thickness of high-efficiency backing layer
1.11	0.02	
0.83	0.03	
0.665	0.04	
0.555	0.05	T: 28% of thickness: transition between the low-efficiency and high-efficiency layers
0.303	0.10	
0.159	0.20	
0.108	0.30	
0.081	0.40	
0.065	0.50	
0.055	0.60	
0.046	0.70	F: 70% of thickness of low-efficiency mat up front
0.040	0.80	
0.037	0.90	
0.033	1.00	

b. It is convenient to assume arbitrarily the thickness of each layer, e.g.,

$$L_F = 21 \text{ mm} \qquad L_T = 8.4 \text{ mm} \qquad L_B = 0.6 \text{ mm}$$

and also assume values of two of the three filtration coefficients, e.g.,

$$\lambda_B = 1.67 \text{ mm}^{-1} \qquad \lambda_F = 0.05 \text{ mm}^{-1}$$

Then the condition

$$1 - \eta_0 = 0.01 = (1 - \eta_F)(1 - \eta_T)(1 - \eta_B) = e^{-\lambda_F L_F - \lambda_T L_T - \lambda_B L_B}$$

can be used to calculate λ_T, giving

$$\lambda_T = 0.30 \text{ mm}^{-1}$$

Next, from

$$\lambda_i = \frac{\eta_{f\beta}(1 - \varepsilon_i)}{D\varepsilon_i} = \frac{0.032}{0.01}\frac{(1 - \varepsilon_i)}{\varepsilon_i}$$

the void fractions can be obtained, i.e.,

$$\varepsilon_F = 0.98 \qquad \varepsilon_T = 0.91 \qquad \varepsilon_B = 0.66$$

The fraction of the dust collected in each layer is calculated next.

$$\eta_F = 1 - e^{-\lambda_F L_F} = 0.650$$
$$= \text{dust fraction collected in front layer}$$
$$\eta_T = 1 - e^{-\lambda_T L_T} = 0.9195$$
$$\therefore (1 - 0.65)0.9195 = 0.3218$$
$$= \text{dust fraction collected in transition layer}$$
$$\eta_B = 1 - e^{-\lambda_B L_B} = 0.6329$$
$$\therefore (1 - 0.65)(1 - 0.9195)0.6329 = 0.0178$$
$$= \text{dust fraction collected in back layer}$$

The corresponding fractional void spaces are

$$\left.\begin{aligned} \varepsilon_F L_F &= 0.98 \times 21 = 20.58 \\ \varepsilon_T L_T &= 0.91 \times 8.4 = 7.644 \\ \varepsilon_B L_B &= 0.66 \times 0.6 = 0.396 \end{aligned}\right\} \qquad \text{Sum} = 28.62$$

$$\therefore \Delta\varepsilon_F = 0.719 \qquad \Delta\varepsilon_T = 0.267 \qquad \Delta\varepsilon_B = 0.0138$$

which are not exactly the same proportions as the fractional dust collections. Iterations are required to find the filter coefficients and the corresponding void fractions which result in exactly identical proportions.

PROBLEMS

1. A bag house filter is run at a constant rate for 20 min, during which time 100 m^3 of gas is filtered. The initial and the final pressure drop over the filter is 2 cm and 10 cm of H_2O column, respectively. It is contemplated to run the filter for one more hour at the final pressure. Calculate the volume of additional gas filtered in this time period.

2. A 10^5 ACFM (at 400°F, 1 atm) gas flow is to be treated in a bag house and is characterized by these values of parameters in the equation

$$\Delta p = v_s(a + bx) \qquad (x \geq x_e)$$

$$a = -25 \text{ kPa s m}^{-1}; \qquad b = 0.15 \text{ kPa s m g}^{-1}$$

$$x_e = 210 \text{ g m}^{-2}; \qquad v_s = 0.05 \text{ m s}^{-1}$$

Estimate

a. the average value of Δp, assuming that Δp is allowed to rise 50% between cleanings;

b. the necessary cloth area A; and

c. the capital and installation cost, by assuming $25/ft^2.

3. A bag filter containing 25 bags, each 2 m^2, has been utilized in constant pressure filtration at 5 cm H_2O column pressure drop. Cleaning time is 35 min.
   a. Justify writing the filtration equation for this filter in the form $t = \text{const } V^2$ (V is m^3 of gas filtered per m^2 of cloth).
   b. Given the value of the constant of 150, estimate the m^3 of gas that can be filtered by this bag house.

4. These are measured data on a bag house operating at constant pressure:

V (m^3/m^2)	82	116	163	232
t (min)	5	10	20	40

   a. What can you say, based on these data, about the resistance of the filter medium compared to that of the filter cake formed by the deposited particles?
   b. If the time of cleaning is 20 min, over what period of filtration time will the maximum amount of gas be filtered per cycle, and what is the volume of gas filtered in this cycle?

5. A gas filtered in a pilot plant scale bag filter gave these results:

	Time t (min)	V (m^3/m^2)
Constant rate period	0–15	4.5
Constant pressure period	15–52	4.5–11.0

   The cleaning time was 10 min. Calculate the maximum volume of gas that can be filtered per day in this filter by using cycles consisting of a constant rate period, followed by a constant pressure period, followed by cleaning, and so forth.

6. A fiber mat air filter is designed to filter 1-μm diameter particles (assumed to be spherical) with a minimum efficiency of 99%. The packing density β of the filter, according to the manufacturer, is 5% and the interstitial velocity of air is designed to be 1 m s^{-1}. The density of the particles is 2 g cm^{-3}.
   a. Calculate the value of $(n_0/D)L$ for the filter.
   b. Using your results from a, calculate
      i. n_0/D
      ii. n_0

for two fiber mat designs:

$$1 \text{ mm thick } (D = 2.3\ \mu\text{m})$$

$$10 \text{ mm thick } (D = 6.8\ \mu\text{m})$$

c. Which of the two mats (1 mm or 10 mm) will become clogged first? Explain.

7. The most important difference between filtering solid particles and mist droplets is that no shaking or other particle removal method has to be provided for the latter because liquids run off the filter fibers. Mist eliminators belong to the fiber mat type of filter category. For very fine mists, mist filters made of very fine fibers are used. So-called fiber demister candles are among the fine mist filters in use. Fiber demister candles are used at superficial velocities of 75–200 mm s^{-1} and pressure drops of 1.25–3.75 kPa, and they collect droplets greater than 5 μm with 100% efficiency and those smaller than 5 μm with 95–99% efficiency. The purpose of this problem is to check and, if possible, verify these claims.

   Let us suppose that the thickness of the fiber mat is $L = 10$ cm, the packing density $\beta = 0.05$, and the overall collection efficiency of the demister candle for 5-μm water droplets is $n_0 = 0.98$. Assuming a superficial velocity of 100 mm s^{-1},

   a. Estimate the individual fiber collection efficiency and the fiber diameter at $t = 20°C$ and 1 atm pressure that must be used for the demister candle to meet the given specifications.

      HINT: Since this calculation involves trial and error, calculate D only to one significant place.

   b. Calculate the overall mist removal efficiency of this demister candle for 1-μm droplets under otherwise identical conditions.

   c. Estimate the pressure drop in kPa over this demister candle under the conditions specified previously.

8. A fibrous mat filter is made of 10 μm-diameter fibers and is to be used to filter 0.5-μm aerosol particles (density, 2.8 g m^{-3}) from air to ambient conditions (assume spherical particles). A face velocity of 1 m s^{-1} is contemplated.

   a. Estimate the individual fiber collection efficiency by all the mechanisms known to you; then form the combined efficiency of an individual fiber.

b. Estimate the permeability of the fibrous mat before use in μm^2 units if the void fraction is 0.8.

c. What is the efficiency of a fibrous mat of $\varepsilon = 0.8$ and $L = 2$ mm? Will this filter have a long life?

d. A fibrous mat filter of 99% efficiency is to be designed from these fibers. The thickness L and the void fractions ε are the design parameters. What is the void fraction for $L = 40$ mm? What is ε for $L = 100$ mm? Which of these two designs will have a longer life (i.e., will plug up later)? Why?

e. A fibrous mat filter of 99% efficiency is to be designed using the previously given fibers. The filter will consist of three layers of equal thickness $L_1 = L_2 = L_3 = 10$ mm. Calculate the void fraction of each layer for the longest life of the filter.

References

Billings, C. E. (1966). Ph.D. thesis, California Institute of Technology, Pasadena.

Carman, P. C. (1956). "The Flow of Gases through Porous Media." Academic Press, New York.

Chen, C. Y. (1955). Filtration of aerosols by fibrous media. *Chem. Rev.* **55**, 595–623.

Kennedy, M. G., and Cheremisinoff, P. N. (1977). Fabric filter dust collectors. *In* "Air Pollution Control and Design Handbook" (P. N. Cheremisinoff and R. A. Youngs, eds.), Part 1, Chap. 11. Dekker, New York.

Collins, R. E. (1961). "The Flow of Fluids through Porous Materials." Reinhold, New York.

Davies, C. N. (1952). Separation of airborne dust and particles. *Proc. Inst. Mech. Eng.* **1B**, 185–198.

Dawson, S. V. (1969). Sc.D. thesis, Harvard University, Cambridge, Massachusetts.

Dorman, R. G. (1964). *In* "High Efficiency Air Filtration" (P. A. White and S. E. Smith, eds.). Butterworth, London.

Dullien, F. A. L. (1979). "Porous Media—Fluid Transport and Pore Structure." Academic Press, New York.

Edmisten, N. G., and Bunyard, F. L. (1970). A systematic procedure for determining the cost of controlling particulate emissions from industrial sources. *J. Air Pollut. Contr. Assoc.* **2**, 446–452.

Fuchs, N. A., and Stechkina, I. B. (1963). A note on the theory of fibrous aerosol filters. *Ann. Occup. Hyg.* **6**, 27–30.

Greenkorn, R. A. (1983). "Flow Phenomena in Porous Media." Dekker, New York.

Happel, J. (1959). Viscous flow relative to arrays of cylinders. *AIChE J.* **5**, 174–177.

Happel, J., and Brenner, H. (1964). "Low Reynolds Number Hydrodynamics." Academic Press, New York.

Iberall, A. S. (1950). Permeability of glass wool and other highly porous media. *J. Res. Natl. Bur. Stand.* **45**, 398–406.

Iinoya, K., and Dennis, R. (1987). Industrial gas filtration. *In* "Filtration" (M. J. Matteson and C. Orr, eds.), Chap. 4. Dekker, New York.

Iinoya, K., and Orr, Jr., C. (1977). Filtration in air pollution. *In* "Air Pollution" (A. C. Stern, ed.), 3rd Ed., Vol. IV, Chap. 4. Academic Press, New York.

Kyan, C. P., Wasan, D. T., and Kintner, P. C. (1970). Flow of single phase fluids through fibrous beds. *Ind. Eng. Chem. Fundam.* **9**, 596–603.

Langmuir, I. (1942). *ORSD Rep.* No. 865.

Matteson, J. M., and Orr, C., eds. (1987). "Filtration—Principles and Practices," 2nd Ed. Dekker, New York.

Pich, J. (1969). Pressure drop in fiber filters at molecular flow. *Staub Reinhalt. Luft.* **29**, 407–408 (in German).

Scheidegger, A. E. (1957). "The Physics of Flow through Porous Media." Univ. of Toronto Press, Toronto.

Strauss, W. (1975). "Industrial Gas Cleaning," 2nd Ed. Pergamon, Oxford.

Sullivan, R. R., and Hertel, K. L. (1940). Flow of air through porous media. *J. Appl. Phys.* **11**, 761–765.

6

Wet Scrubbers

6.1 Introduction

Wet scrubbers are used to remove either gaseous or particulate contaminants from effluent gas streams. Applications include use in industrial boilers, lime kilns, foundry cupolas, municipal sludge incinerators, and odor control devices. The wet scrubber used for the removal of gaseous contaminants is usually an absorption tower or a venturi scrubber. There is usually particle removal along with gas absorption, which is discussed separately in Chapter 8.

Wet collection of particulates has many different features when compared with the methods discussed previously, such as gravity, momentum and centrifugal separators, and filters, which are dry methods for particulate collection. Broadly speaking, in wet scrubbers the particles are collected either by liquid drops or a continuum of liquid in the form of wetted walls, liquid sheets, etc. It is important that the liquid should wet the particles. Wet collection of particles has a number of advantages over dry methods, such as reduced dust explosion risk and being able to handle hot gases, sticky particulates, and liquids, as well as a number of disadvantages, largely associated with the problems of effluent liquid disposal and removal or recycling of the contaminants from a liquid dispersion. For scrubber applications, maintenance, and control see Calvert *et al.*, 1972.

The principal collection mechanisms in scrubbers are the same as in filters: inertial impaction, interception, and diffusion. In addition, condensation effects and particle entrainment may also play important roles in liquid scrubbers.

There are a large variety of different types of scrubbers in use, the most important of which are spray towers and chambers, packed and plate columns, impingement type scrubbers, and finally, the venturi scrubbers. First, the simplest, least expensive and lowest energy scrubber, the simple spray tower, is described.

6.2 Spray Towers

Operating Conditions

In a counter-current spray tower, drops of liquid are introduced by spray nozzles and are allowed to fall downward through a rising stream of dirty gas. In order not to be entrained by the gas stream, the droplets must be large enough to have a falling speed greater than the upward velocity of the gas stream, which is usually in the range of 0.6–1 m s^{-1}.

Droplets smaller than 1 mm tend to be approximately spherical, and their falling speeds can be estimated by known methods. The droplets used in spray towers are on the order of 0.1–1 mm in diameter. Under these conditions, the predominant collection mechanisms are inertial impaction and interception. Collection efficiency increases with the relative velocity between the particle and the droplet, and it decreases with the droplet size. Smaller droplets, however, fall slower, and therefore, there is an optimum drop size for the most efficient collection of particles of a given size, as shown in Fig. 6.1.

Typical large spray towers are shown in Fig. 6.2. The design shown on the left [(a)] is 5 m in diameter, has an effective height of 16 m, and has 5

Figure 6.1 Optimum droplet size for collection by inertial impaction by droplets falling in the earth's gravitational field, that is, in a simple spray tower (Stairmand, 1950).

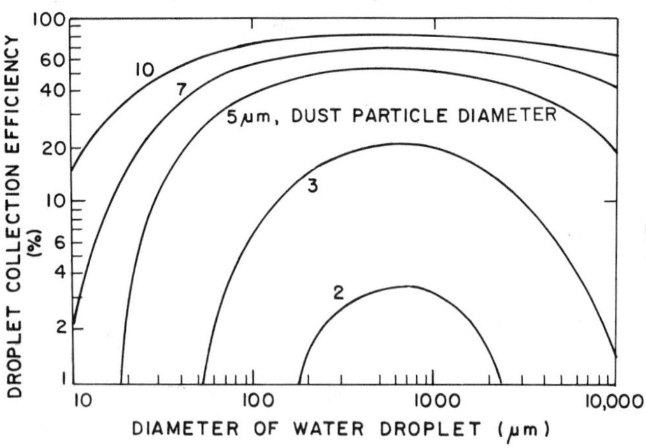

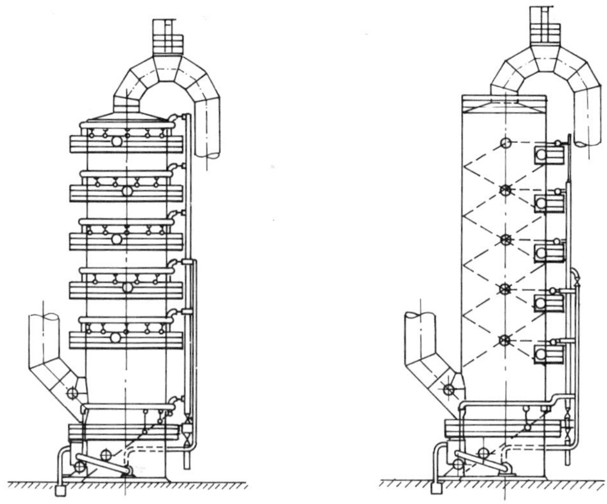

Figure 6.2 (a) Spray tower with circumferentially placed sprays. (b) Spray tower with axially placed sprays (Hausberg, 1961).

circumferentially arranged rings of sprays with 14–16 spray nozzles per ring. The tower handles 50,000 m^3 of gas per hour, and the gas velocity is about 0.75 m s^{-1}. The arrangement on the right [(b)] shows the same tower with nine centrally placed solid cone impact sprays, which resulted in an increase of efficiency to 90% from 73% on particles greater than 5 μm. Typical pressure drops are in the range of 2–4 in. w.c. Several sections, each with one set of spray nozzles, are required, because in the case of hot gases, the droplets gradually evaporate while falling, whereas in the absence of evaporation, the droplets coagulate and form fewer and larger drops, and this results in a decreased dust removal efficiency. The specific water consumption W_{sp} is an important parameter in scrubbers. It gives the number of liters (cubic decimeters) of water used per cubic meter of gas treated. The overall efficiency of a spray tower increases with increasing specific water consumption. A correlation between specific water consumption and the efficiency of a simple spray tower used to remove blast furnace dust is shown in Fig. 6.3.

Overall Efficiency of Particle Collection in Spray Towers

The overall mechanism of particle collection in spray towers can be understood from a simple analysis, following lines of thought similar to

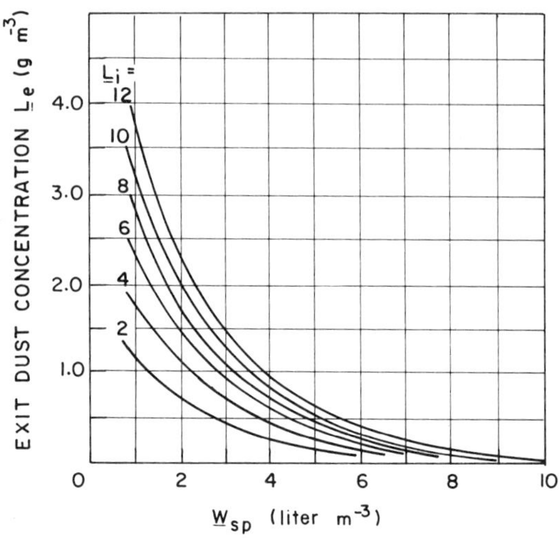

Figure 6.3 Specific water consumption correlation for a simple spray tower (Hausberg, 1961): L_i, dust concentration at inlet (g m^{-3}); L_e, dust concentration at exit; W_{sp}, specific water consumption.

those used in Chapter 5, in the case of fibrous filters. As usual, it is assumed that the aerosol particles (dust) or droplets have the same velocity as does air. The analysis is greatly simplified by the assumption of a uniform and constant droplet diameter, D. In reality, there is a distribution of droplet diameters, which keep decreasing because of evaporation or increasing because of coagulation. By this assumption, water droplets fall through the aerosol; and their relative velocity with respect to the gas, and also with respect to the dust particles, is equal to u_t, the terminal settling velocity of the droplets. This is also true when the air is flowing countercurrent to the water droplets.

Consider the schematic diagram (Scheme 6.1), showing the cross section of the spray tower.

$$\mathbf{v}_L - \mathbf{v}_G = \mathbf{u}_t = \text{velocity of spray relative to dusty gas}$$

that is,

$$v_L + v_G = u_t \tag{6.1}$$

The thickness dh should be thought of as greater than a drop diameter, but not great enough to allow the projection of the drops onto the plane of tower cross section to overlap; that is, it corresponds to a "monolayer" of drops, in the statistical average sense of the word.

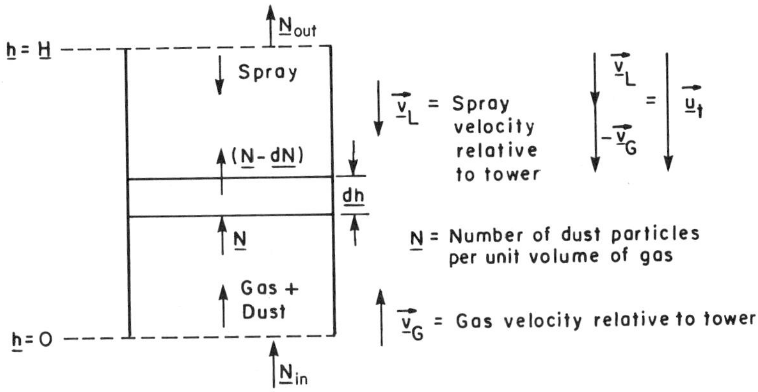

Scheme 6.1

In order to simplify the problem conceptually, let us first consider the special case of $v_L = 0$ (i.e., $\mathbf{v}_G = -\mathbf{u}_t$). In this case, the position of the spray drops (consider uniform drop size, D, for simplicity) is fixed relative to the tower, and the situation is analogous to the case of the fibrous filter: a fixed "bed" of droplets is traversed by a gas of velocity relative to the drops equal to u_t. In this case, the position of the volume element of thickness dh is also fixed relative to the tower. Usually, $v_G < u_t$, that is, the spray is allowed to settle with a velocity of $v_L = u_t - v_G$; but relative to the aerosol (gas + dust), the situation is the same as in the case of $v_L = 0$, except that the slice of thickness dh now travels downward in the tower with a steady velocity v_L. The volume of the slice is $dh\,A_T$, where A_T is the cross-sectional area of the tower. The total projected area of drops per unit volume of mixture (gas + liquid) is equal to number of drops per unit volume of mixture (gas + liquid) times $D^2\pi/4$. The number of drops per unit volume of mixture (gas + liquid) is, in an excellent approximation, given by dividing the volume of water present per unit volume of gas by the volume of a drop:

$$\frac{c_w\ (\mathrm{kg\ H_2O/m^3\ gas})}{\rho_w\ (\mathrm{kg\ H_2O/m^3\ H_2O})} \bigg/ \frac{D^3\pi}{6} \tag{6.2}$$

as the volume of water is very small compared to the gas volume. Here, c_w is the mass of water per unit volume of gas. Therefore, the total projected area of drops in the volume element is

$$\frac{c_w}{\rho_w}\frac{6}{D^3\pi}\frac{D^2\pi}{4}\,dh\,A_T = \frac{3}{2}\frac{c_w}{\rho_w}\frac{A_T}{D}\,dh \tag{6.3}$$

Under the conditions stated, one can apply the single droplet collection efficiency concept to the drops in the slice, that is, one lets

$$\eta = \frac{\text{number of dust particles captured in slice per unit time}}{\text{number of dust particles crossing the projected area of drops per unit time}}$$

$$\therefore \eta = \frac{-dN\,u_t A_T}{\dfrac{3}{2}\dfrac{c_w}{\rho_w}\dfrac{A_T}{D}\,dh\,u_t N} \tag{6.4}$$

where $u_t A_T$ is the volume flow of dusty gas relative to the slice.

Imagine a "snap-shot" taken of the contents of the tower, showing N as a function of h in a steady state. This functional dependence can be calculated by integrating Eq. (6.4) from the bottom of the tower, where N has the known value N_{in}, to any height h. We are particularly interested in the exit dust concentration, and therefore, we integrate to the top of the column, that is, $h = H$:

$$\int_{N_{in}}^{N_{out}} \frac{dN}{N} = -\frac{3}{2}\frac{\eta}{D}\frac{c_w}{\rho_w}\int_0^H dh \tag{6.5}$$

giving the result

$$\ln\frac{N_{out}}{N_{in}} = -\frac{3}{2}\frac{\eta}{D}\frac{c_w}{\rho_w}H \tag{6.6}$$

Comparing Eq. (6.6) with Eq. (5.28), we see that apart from different numerical coefficients the two equations are identical.

The ratio c_w/ρ_w is a very good approximation of the volume fraction occupied by the water in the tower, that is, $1 - \varepsilon$. As $c_w/\rho_w \ll 1$, it follows that in the spray tower $\varepsilon \approx 1$.

Next, we replace c_w/ρ_w with an expression containing the gas and liquid flow rates. Assuming that the droplet concentration in the tower is uniform, it can be shown that the area fraction A_L/A_T of water drops in the tower cross section A_T is equal to the volume fraction, that is,

$$c_w/\rho_w = A_L/A_T \tag{6.7}$$

where A_L is the sum of the cross-sectional areas of water drops in the cross section of the tower, A_T.

However, A_L is related to the throughput of water Q_L and the droplet velocity v_L as

$$A_L = Q_L/v_L \tag{6.8}$$

By Eq. (6.1),

$$v_L = u_t - v_G = u_t - Q_G/A_T \tag{6.9}$$

where Q_G is the gas throughput through the tower. (It is realized that A_L is very much smaller than A_T and therefore it can be neglected.)

Combining Eqs. (6.7)–(6.9) gives this result[1]:

$$\frac{c_w}{\rho_w} = \frac{Q_L/A_T}{u_t - Q_G/A_T} \ll 1 \tag{6.10}$$

Inserting this expression in Eq. (6.6), after exponentiating and using the relationship between the overall efficiency of the tower η_0 and the ratio N_{out}/N_{in}, this relationship is obtained for η_0:

$$\begin{aligned}\eta_0 &= 1 - \exp\left(-\frac{3}{2}\frac{\eta}{D}H\frac{Q_L/A_T}{u_t - Q_G/A_T}\right) \\ &= 1 - \exp\left(-\frac{3}{2}\frac{\eta}{D}H\frac{v_G}{u_t - v_G}\frac{Q_L}{Q_G}\right)\end{aligned} \tag{6.11}$$

In the special case when the velocity of the liquid relative to the tower is equal to the gas velocity relative to the tower, that is,

$$v_L = v_G \tag{6.12}$$

we have

$$v_L = Q_G/A_T \tag{6.13}$$

and

$$u_t - v_G = v_L \tag{6.14}$$

Therefore, in this case, Eq. (6.11) simplifies to

$$\eta_0 = 1 - \exp\left(-\frac{3}{2}\frac{\eta}{D}\frac{Q_L}{Q_G}H\right) \tag{6.15}$$

Equation (6.15) is particularly instructive, showing the dependence of the overall efficiency of a spray tower on the various parameters for this special case. Evidently, the overall efficiency increases exponentially with the ratio of the individual drop collection efficiency to the drop diameter, tower height, and liquid-to-gas flow rate ratio in the tower.[2] Because of the simplifying assumptions made in the derivation of Eq. (6.11), this equation cannot be used for accurately designing spray towers, but it shows correctly the trends of dependence of the tower's collection efficiency on some of the most important parameters. It is worth pointing out that since the individ-

[1] To keep c_w/ρ_w constant, Q_L is the maximum for $Q_G = 0$; as Q_G is increased, Q_L must be decreased (i.e., Q_L/Q_G is decreased even more). Finally, in the limit as $Q_G/A_T \rightarrow u_t$, $Q_L \rightarrow 0$.

[2] It is noted that the condition $v_L = v_G$ fixes the value of Q_G, but Q_L can be chosen arbitrarily as long as $Q_L/Q_G \ll 1$.

ual drop collection efficiency η is a function of the drop diameter D, the two quantities are not independent of each other and the ratio η/D is best regarded as a single parameter.

Spray chambers in which the particle laden air flows horizontally through a vertical settling spray would be completely analogous to the case of a fibrous filter, were the drops not blown by the air, resulting in a vanishing relative velocity between the drops and the particle after a relatively short contact time. As a result, drops are effective incapturing particles only near the top of the chamber where they enter. For this cross-flow type of scrubber to be effective, the water must be introduced at a high velocity and a high flow rate. Settling drops are quite ineffective.

6.3 Impactor Scrubbers

Plate and packed bed scrubbers belong to this class of scrubbers (Calvert, 1977b). They are medium-energy scrubbers, with pressure drops in the range of 3–8 in. w.c. The plate column scrubbers comprise sieve (perforated) and impingement (Peabody-type) plates. The latter is shown diagrammatically in Fig. 6.4 (Ray, 1952).

It is customary to use the penetration $Pt = (1 - \eta_0)$, which is equal to $N_{\text{out}}/N_{\text{in}}$, instead of the overall efficiency η_0. This quantity is more convenient in some respects than the efficiency. For example, it is more concise to say that the penetration of a particular scrubber has been reduced by a factor of 2 than that the efficiency has been increased from $1 - x$ to $1 - x/2$.

The following empirical correlation has been developed to predict the collection efficiency of hydrophilic particles by water in a sieve plate scrubber (Taheri and Calvert, 1968):

$$Pt = \exp(-40F^2 v_{\text{Gh}} d_{\text{a}}^2/9\mu D_{\text{h}}) \tag{6.16}$$

Here, v_{Gh} is the gas velocity in the holes of a sieve plate, D_{h} is the hole diameter, and

$$d_{\text{a}} = d(\rho_{\text{p}} C)^{1/2} \tag{6.17}$$

is a quantity, usually referred to as the *aerodynamic impaction diameter* of a particle; d_{a} contains all the particle properties of importance contained in the inertial impaction parameter Ψ ($2\Psi = v_{\text{Gh}} d_{\text{a}}^2/9\mu D_{\text{h}}$). Obviously the dimensions of d_{a} are not length, and therefore, the term diameter is, physically speaking, wrong. Finally, F in Eq. (6.16) is a plate efficiency. As Ψ is dimensionless, consistent units must be used in its calculation.

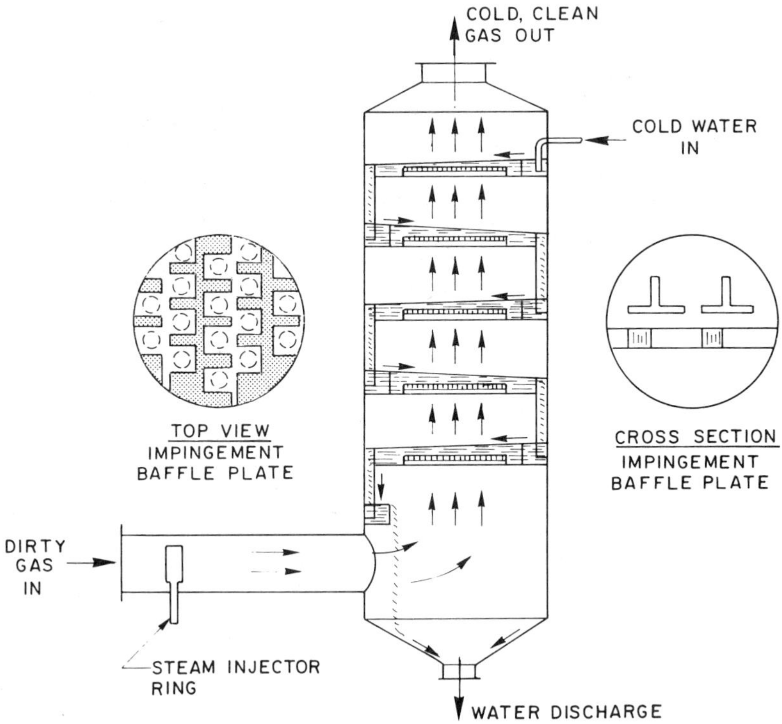

Figure 6.4 Impingement plate scrubber (called the Peabody scrubber) (Ray, 1952).

By setting *Pt* equal to 0.5, Eq. (6.16) may be solved for the aerodynamic impaction cut diameter. The aerodynamic impaction cut diameter is plotted in Fig. 6.5 for typical sieve plate conditions against the hole velocity and the hole diameter. For impingement plates, predicted aerodynamic impaction cut diameters have been plotted in Fig. 6.6.

For a packed column, the penetration has been correlated by the equation (Jackson and Calvert, 1966)

$$Pt = \exp(-7.0 H v_s d_a^2 / 9 \mu \varepsilon D_c^2) \tag{6.18}$$

In this equation, v_s is the usual superficial gas velocity and D_c is the diameter of the packing. Aerodynamic cut diameters calculated by Eq. (6.18) are shown in Fig. 6.7. Consistent units must be used for all the quantities in the dimensionless exponent of Eq. (6.18).

Packed beds are seldom used for particle collection because the particles have a tendency to settle out in the passages of the packing and gradually clog up the entire bed. A notable exception is the case of soluble particles.

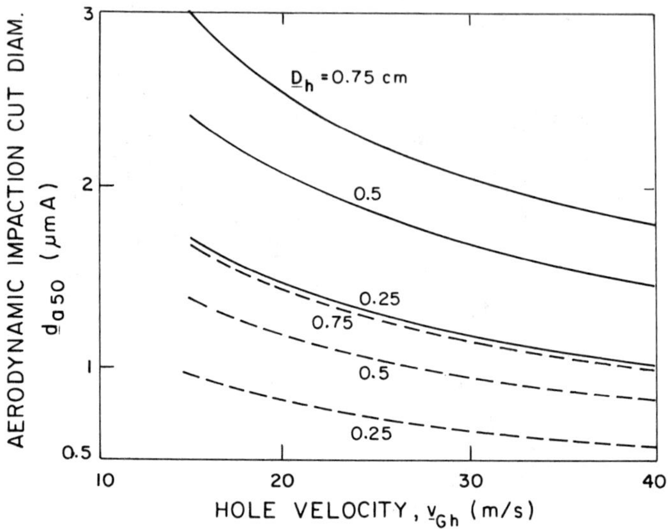

Figure 6.5 Performance cut diameter prediction for typical sieve plate conditions. Solid line, $F = 0.4$; dashed line, $F = 0.65$ (Taheri and Calvert, 1968).

Figure 6.6 Performance cut diameter predictions for typical impingement plate conditions (Taheri and Calvert, 1968).

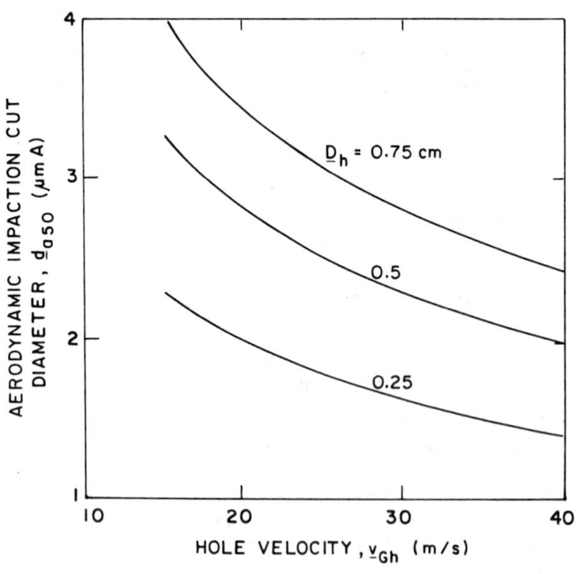

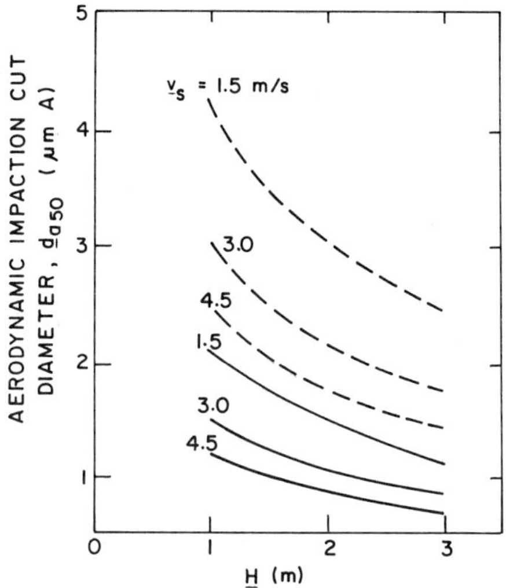

Figure 6.7 Performance cut diameter predictions for typical packed bed conditions. Solid line, $D_c = 2.5$ cm; dashed line, $D_c = 5.0$ cm; $\varepsilon = 0.75$ (Jackson and Calvert, 1966).

Figure 6.8 Floating bed scrubber. (Courtesy of Aerotec Industries, Inc., South Norwalk, Connecticut.)

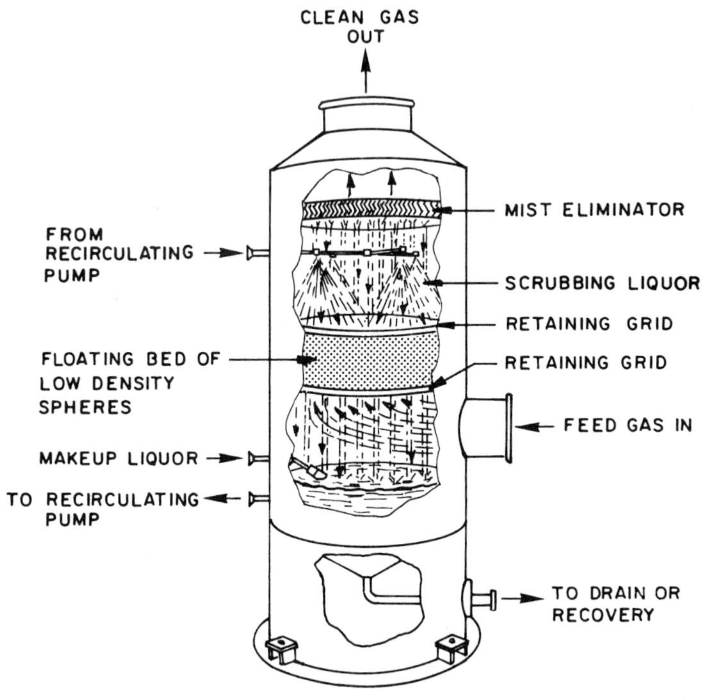

A modified unit, called the *floating bed scrubber*, has been developed. Here, low-density spheres (Ping-Pong balls) are contained between grids, and the bed tends to float on the rising gases. The fluidized state of the bed prevents plugging and so avoids the chief problem of operating a packed bed scrubber. A floating bed scrubber is shown in Fig. 6.8.

Scrubbers with Self-Induced Spray or Bath-Type Scrubbers

In this type of scrubber, the droplets are formed by the gas stream breaking through a sheet of liquid or impinging on a pool. The best known scrubbers in this category are the Rotoclone N scrubber, shown in Fig. 6.9, and the Doyle scrubber, shown in Fig. 6.10. These units are extensively used in the metallurgical industry for collecting coarse dusts and sticky materials as

Figure 6.9 Self-induced spray-type scrubber (called the Rotoclone N scrubber). (Courtesy of American Air Filter Co. Inc., Louisville, Kentucky.)

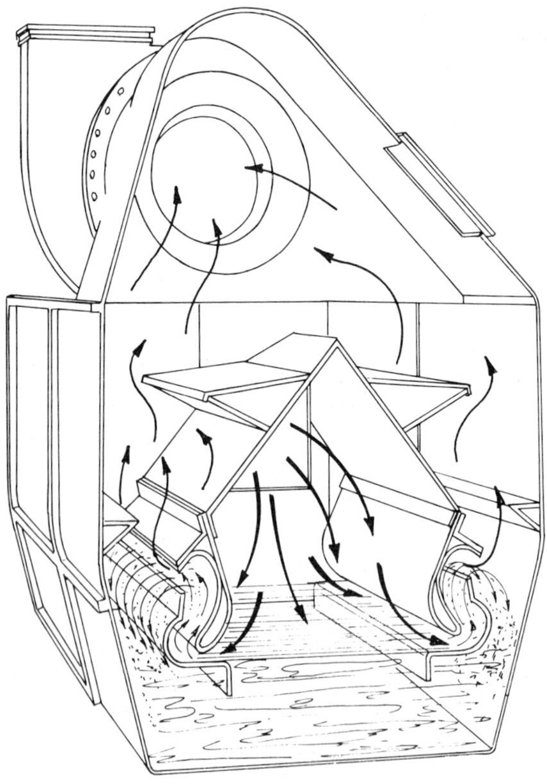

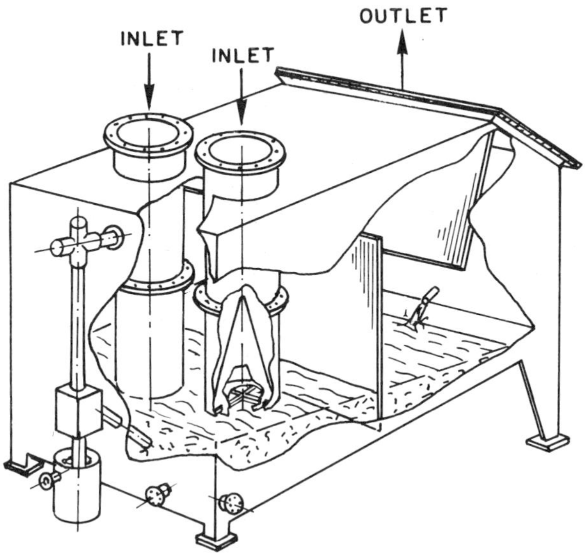

Figure 6.10 Self-induced spray-type scrubber (called the Doyle scrubber) (Doyle and Brooks, 1957).

well as for the collection of explosive dusts. They are also medium-energy scrububers with power requirements between 0.5 and 1.3 kw/1000 m^3 per hour of gas flow (about 7–8 in w.c.).

6.4 Wet Dynamic and Cyclonic Scrubbers

Scrubbers of a variety of different designs utilize centrifugal force. Cyclone spray scrubbers consist of a vertical cylindrical body at the axis of which hydraulically atomized spray nozzles spray liquid radially. The gas enters tangentially through a tapered inlet duct, so that an increasing gas velocity is produced (Schifftner and Hesketh, 1983).

Wet dynamic scrubbers are based on a centrifugal fan or blower in which intake water is sprayed by means of a spray nozzle. A typical centrifugal fan wet scrubber is the Rotoclone W. The view is fairly commonly held that improved scrubber efficiency depends on increased centrifugal force, because the chief collection mechanism is assumed to be inertial impaction. It has been demonstrated that mists and even fumes will coagulate, and at the same time capture dust, in a relatively slow turning (600–1500 rpm) centrifugal fan where the centrifugal force is insufficient for the removal of

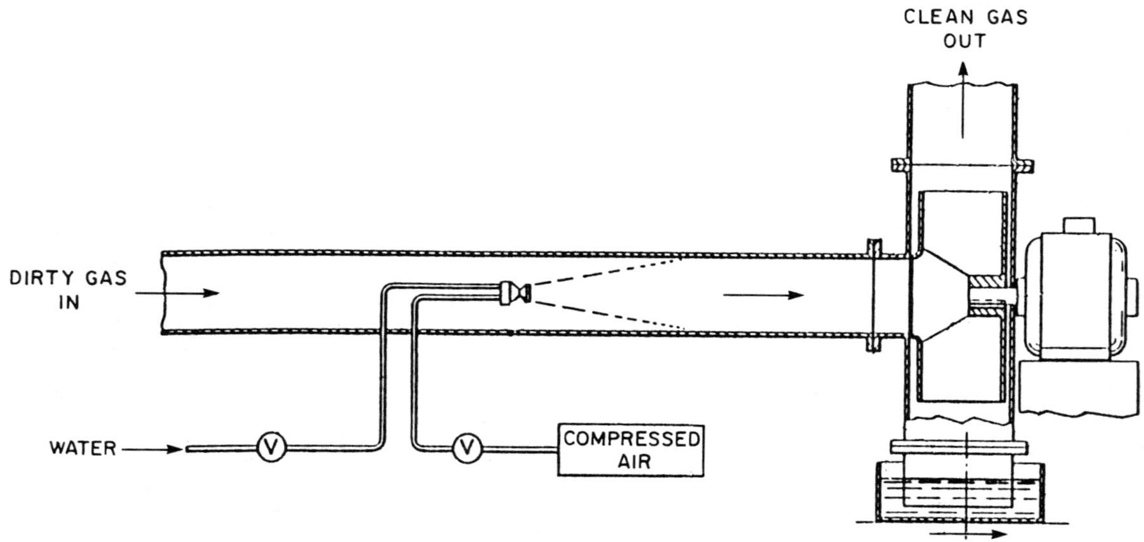

Figure 6.11 Schematic diagram of the Waterloo scrubber (Dullien and Spink, 1978).

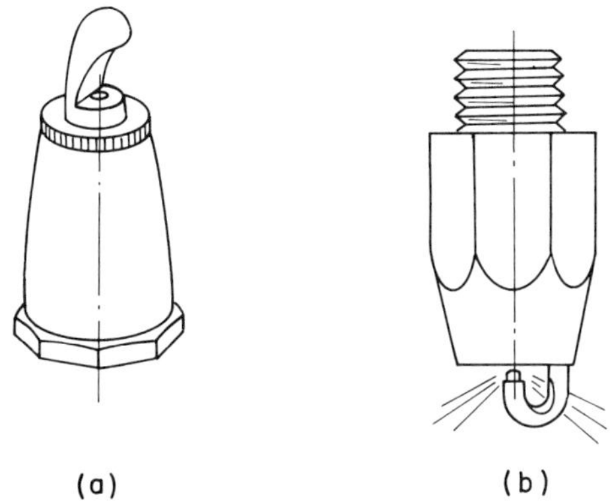

Figure 6.12 Impingement-type nozzles. (a) Fixed wall—directional spray. (b) Hook type—hollow cone spray (Marshall, 1954).

such finely dispersed aerosols. The coagulation is presumed to take place as a result of a synergistic effect involving inertia and turbulence (Dullien, 1977). The Waterloo scrubber, a low-to-medium-energy, high-efficiency scrubber, shown diagrammatically in Fig. 6.11, takes advantage of this effect, which is similar to the formation of rain drops, by using a very fine spray in the fan intake (Dullien and Spink, 1978).

Figure 6.13 Swirl-type spray nozzles. Nozzle designs with swirl chamber (Marshall, 1954).

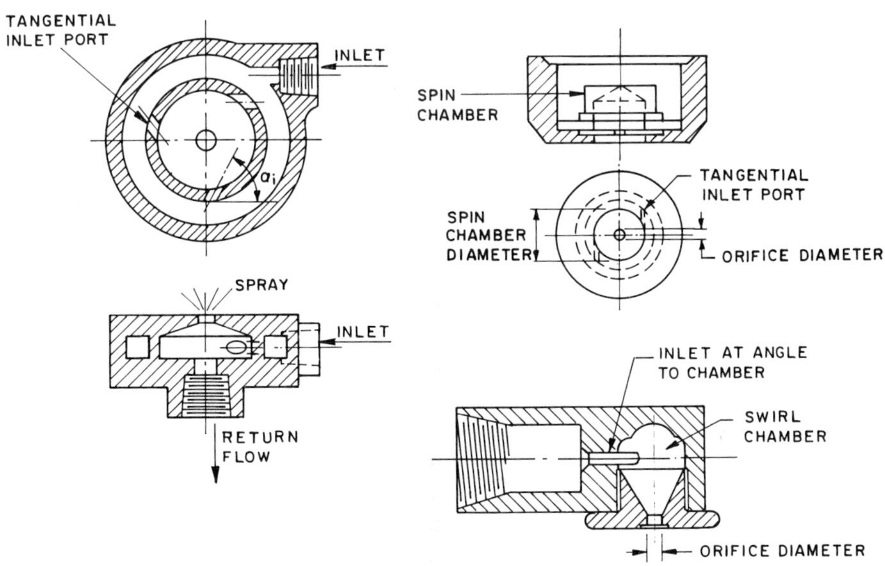

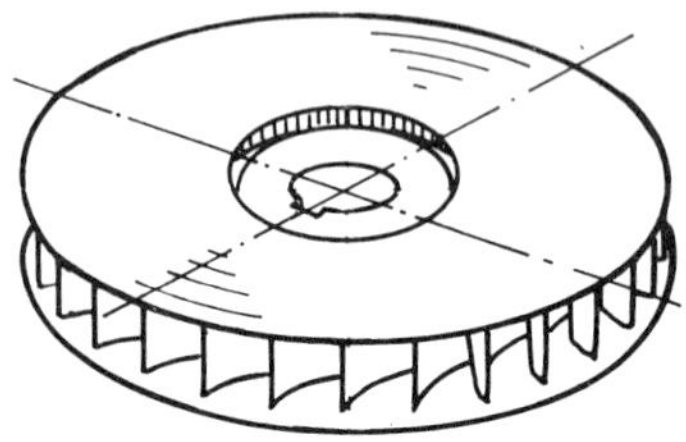

Figure 6.14 Typical disc in spinning disc atomizers (Marshall, 1954).

6.5 Spray Nozzles

Many different types of scrubbers need spray nozzles for their operation. The nozzle characteristics are important for the efficient and economical operation of the scrubber. As the scrubbing liquid often contains suspended solid particles, it is very important that the nozzle neither plugs nor erodes as a result of the presence of solids Another requirement is the generation of fairly uniform and often very fine spray at the cost of relatively little energy. Atomization of a liquid by any spray nozzle available today is energetically very inefficient; only approximately 1% of the energy applied is used for atomizing.

The nozzle types of importance include the following:

1. Impingement type nozzles [see Figs. 6.12(a) and (b)]
2. Spiral-type spray nozzles (see Fig. 6.13)
3. Spinning disc-type nozzles (see Fig. 6.14)
4. Two-liquid jet impingement nozzles

Figure 6.15 Pneumatic or two-fluid atomizing nozzle. (Courtesy of Caldyn Apparatebau GmbH, Federal Republic of Germany.)

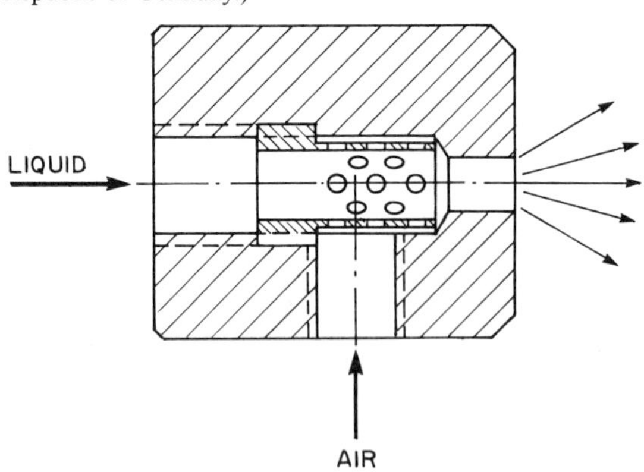

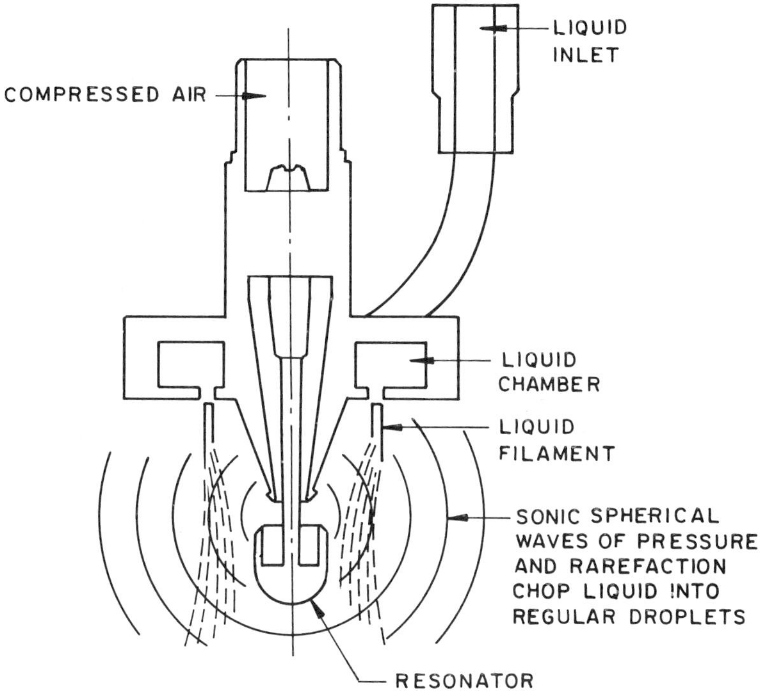

Figure 6.16 Sonic spray nozzle (Anonymous, 1961).

5. Pneumatic or two-fluid atomization nozzles (see Fig. 6.15)
6. Sonic spray nozzles (see Fig. 6.16)

6.6 Venturi-Type Scrubbers

Operating Conditions

Examples for conventional venturi scrubbers are shown diagrammatically in Fig. 6.17 (Jones and Anthony, 1952). As seen in the figure, water is injected into the throat of a venturi, where the velocity of the gas is highest. The gas, while accelerating the injected water to the gas velocity, breaks up the water into a relatively fine spray. The particles are captured by the water droplets primarily by inertial impaction. Because of the high relative velocity in the venturi throat between the gas and the water and because of the high specific water consumption of the venturi, very high collection

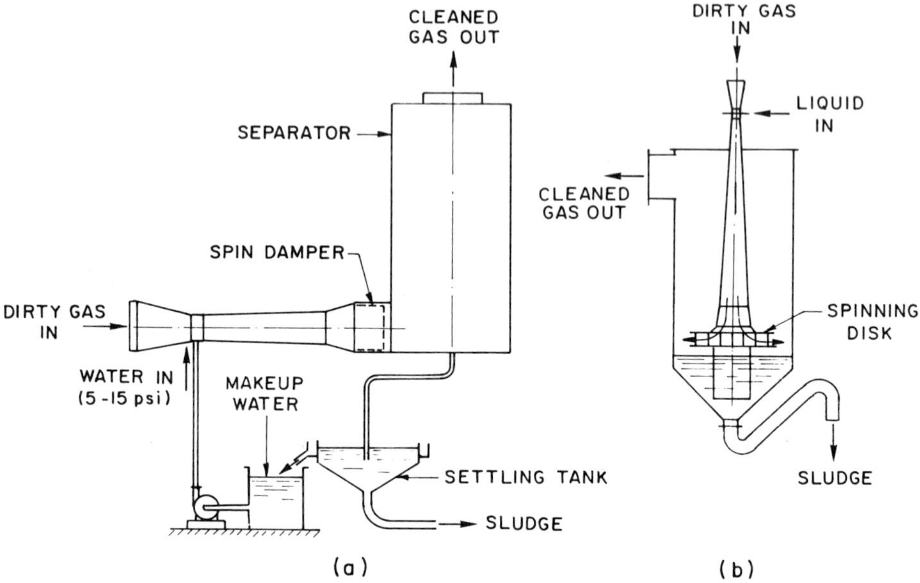

Figure 6.17 Pease–Anthony-type venturi scrubber. (a) Tangential entry to separating chamber. (b) Turbine entry to separating chamber (Jones and Anthony, 1952).

Figure 6.18 Predicted penetration for venturi scrubber (Calvert *et al.*, 1972).

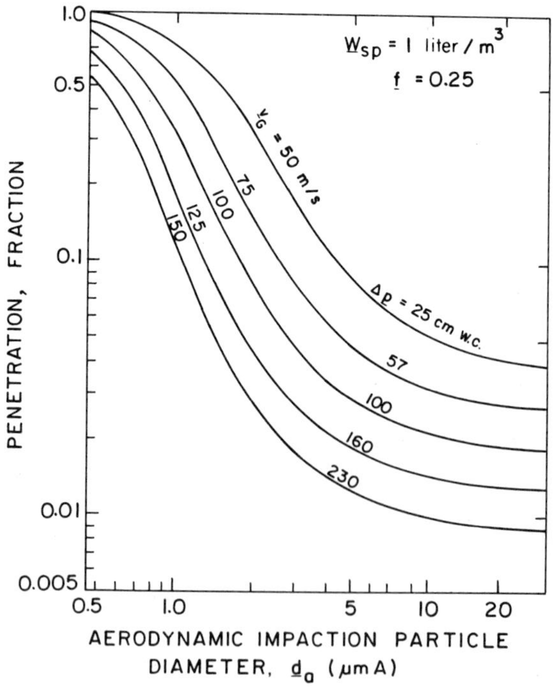

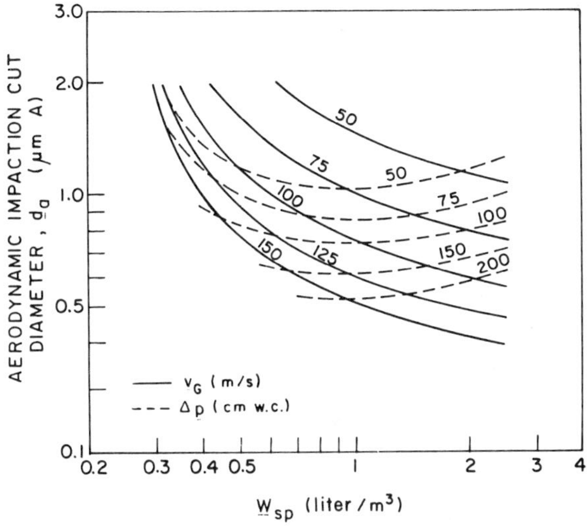

Figure 6.19 Performance cut diameter predictions for venturi scrubber; $f = 0.25$ (Calvert *et al.*, 1972).

efficiencies, accompanied by high pressure losses, have been reported (see Figs. 6.18 and 6.19).

Empirical Penetration Formulas

An empirical relationship that is used to estimate the penetration in a venturi (Calvert *et al.*, 1972) is

$$Pt = \exp\left[\frac{2Q_L v_G \rho_w D}{55 Q_G \mu_G} F(\Psi, f)\right] \tag{6.19}$$

where Ψ is evaluated at the gas velocity, v_G, in the throat, μ_G is the gas viscosity, and ρ_w is the water density.

$$F = \frac{1}{2\Psi}\left[-0.7 - 2\Psi f + 1.4\ln\left(\frac{2\Psi f + 0.7}{0.7}\right) + \left(\frac{0.49}{0.7 + 2\Psi f}\right)\right] \tag{6.20}$$

The empirical factor f is usually between 0.1 and 0.5. The dimensionless exponent in Eq. (6.19) must be evaluated, using consistent units for all quantities.

The drop diameter D to be used in Eqs. (6.19) and (6.20), and for venturi scrubbers in general, is the Sauter surface mean diameter, which is computed from the correlation obtained for pneumatic atomization (Nukiyama and Tanasawa, 1938):

$$D = \frac{58{,}600}{v_G}\left(\frac{\sigma}{\rho_L}\right)^{1/2} + 597\left(\frac{\mu_L}{\sqrt{\rho_L \sigma}}\right)^{0.45}\left(1{,}000\frac{Q_L}{Q_G}\right)^{1.5} \tag{6.21}$$

In this equation, consistent cgs units are used to give D in μm; σ is the surface tension of water, and μ_L and ρ_L are the viscosity and density of water, respectively.

A simpler and older equation used for predicting penetration in a venturi scrubber was given by Johnstone *et al.* (1954):

$$Pt = \exp\left(-k\frac{Q_L}{Q_G}\sqrt{\Psi}\right) \tag{6.22}$$

where Ψ is evaluated at the gas velocity, v_G, in the throat, and if consistent units are used, the value of k is usually between 13 and 27.

Typical results for the penetration in venturi scrubbers as a function of the aerodynamic impaction diameter is shown in Fig. (6.18) (Calvert *et al.*, 1972). The gas velocity in the throat and the pressure loss are also shown as parameters. A correlation showing the aerodynamic cut diameter as a function of the specific water consumption, gas velocity, and pressure loss is shown in Fig. 6.19 (Calvert *et al.*, 1972). Other correlations, showing the collection efficiency for venturi scrubbers as a function of the same parameters, are shown in Fig. 6.20 (Johnstone and Eckman, 1951).

Pressure Loss in Venturi Scrubbers

The main source of pressure loss in venturi scrubbers is the mechanical energy used for accelerating the water injected into the venturi to the gas velocity in the throat. If the water is injected into the throat, a mechanical energy balance over the throat gives this expression for the pressure loss:

$$\Delta p = p_1 - p_2 = \tfrac{1}{2}v_G^2 \rho_L \frac{Q_L}{Q_G} \tag{6.23}$$

where, if consistent SI units are used, Δp is obtained in pascals (for derivation, see Example 6.4). One of the assumptions made in the derivation of Eq. (6.23) is that there is no evaporation of water. If some water evaporates, the volume flow and the gas velocity at the exit of the throat

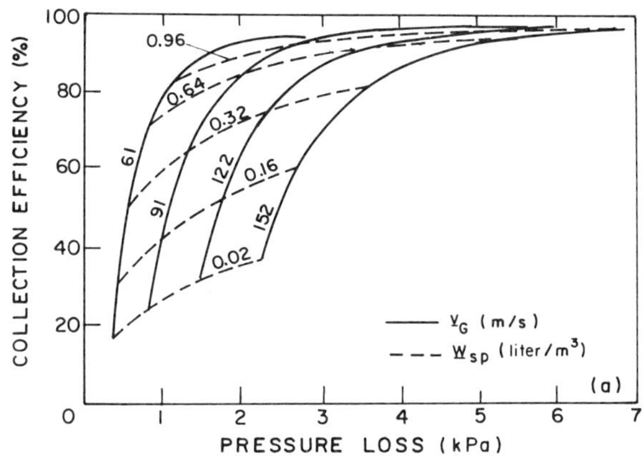

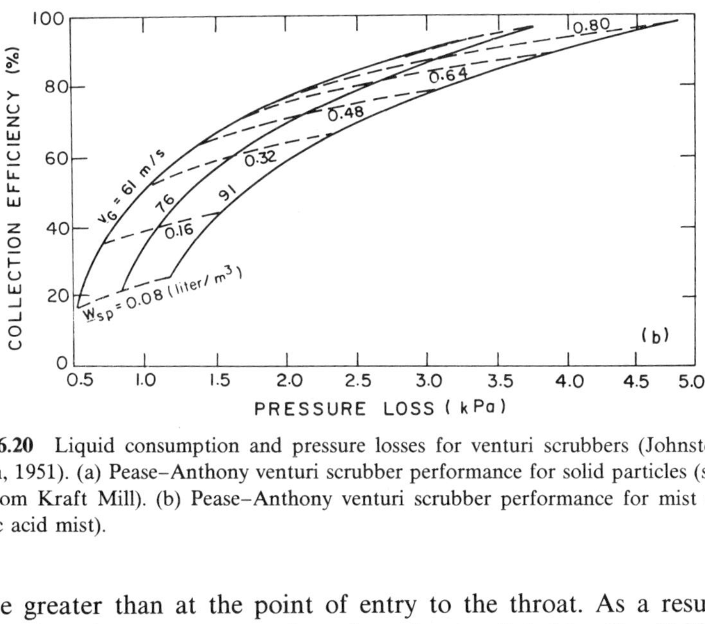

Figure 6.20 Liquid consumption and pressure losses for venturi scrubbers (Johnstone and Eckman, 1951). (a) Pease–Anthony venturi scrubber performance for solid particles (salt cake fume from Kraft Mill). (b) Pease–Anthony venturi scrubber performance for mist droplets (sulfuric acid mist).

become greater than at the point of entry to the throat. As a result, the pressure drop becomes greater than the value predicted by Eq. (6.23) by a factor usually between 1 and 2 (Doyle and Brooks, 1957). This can be verified by reading the values of Δp, v_G, and Q_L/Q_G from Fig. 6.19 and substituting these into Eq. (6.23).

An empirical equation given by Hesketh (1974) is

$$\Delta p = 190 v_G^2 \rho_G A_t^{0.133} (Q_L/Q_G)^{0.78} \tag{6.24}$$

where, if consistent SI units are used, Δp is obtained in pascals (A_t is the throat cross-sectional area).

Present day venturi scrubbers can be quite varied in configuration. In some scrubbers, the throat opening can be varied by movable plates, annular discs, etc. There are rod-type venturi scrubbers in which the throat area reduction is achieved by placing bundles of rods in a chamber having straight sides. In these, the gas and the liquid are in countercurrent flow [see Fig. 6.21; Hesketh (1981)].

Figure 6.21 Rod-type venturi scrubber with adjustable throat size (Hesketh, 1979).

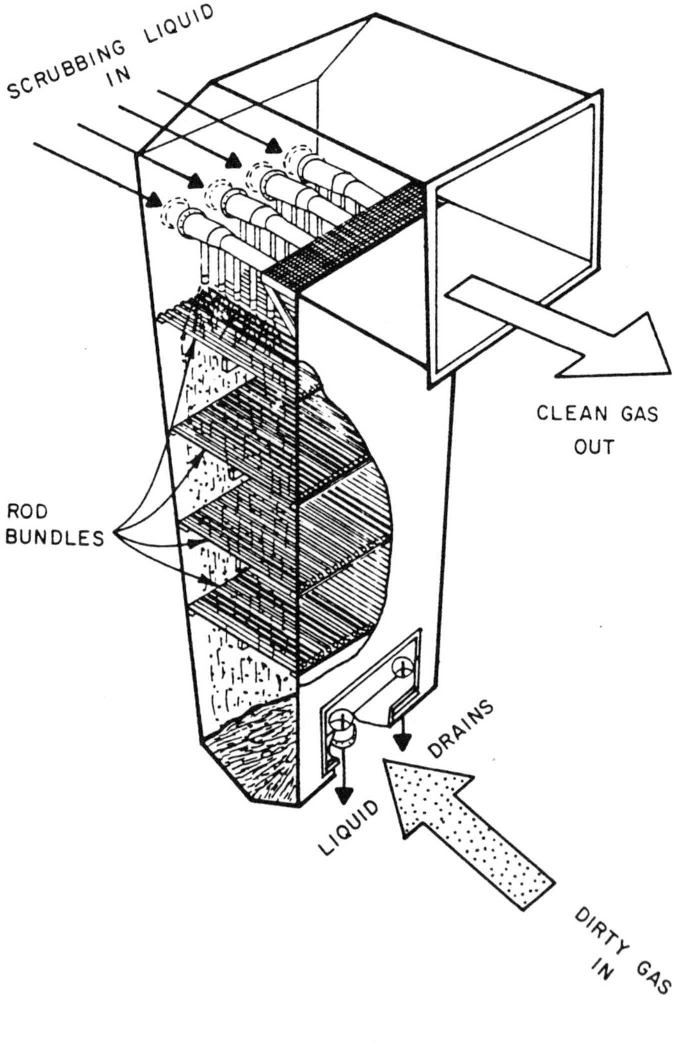

6.7 Correlation of Scrubber Efficiency and Pressure Loss

In general, for one and the same dust, a higher collection efficiency requires greater power consumption; in order to obtain the same efficiency of collection of a finer dust, more power is needed. These quantities have been correlated on a semiquantitative basis by several researchers, but particularly thoroughly by Semrau (1960).

The pressure loss, Δp, expressed in any arbitrary units, for example, in kilopascals, is equivalent to the mechanical energy dissipated per unit volume of fluid (for example, kW h/1000 m^3) or power used per unit volume flow rate of fluid (for example, kW/1000 m^3 h^{-1}). These latter two are denoted by the same symbol P (for power) as they are evidently equal. The conversion from Δp to P is obtained as shown here:

kPa 10^3 Pa	kg m m^{-2} s^{-2}	kg m^2 m^{-3} s^{-2}	J m^{-3}	W s m^{-3}	kW s m^{-3}	kW h m^{-3}	10^3 kW h/1000 m^3
kPa	Pa	kg m m^{-2} s^{-2}	kg m^2 m^{-3} s^{-2}	J m^{-3}	10^3 W s m^{-3}	3600 kW s m^{-3}	kW h m^{-3}

$$\therefore 1 \text{ kPa} = 0.2778 \text{ kW h/1000 m}^3$$

Figure 6.22 Correlation of scrubber efficiency, measured in transfer units, to contacting power for a series of aerosols, independent of scrubber type (Semrau, 1960). FCE, Furnace.

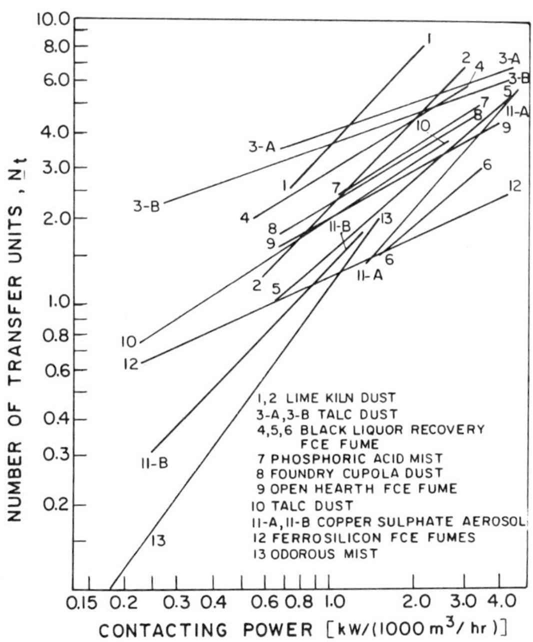

Thus,

$$\Delta p\ (\text{kPa}) \times 0.2778\ \frac{\text{kW h}/1000\ \text{m}^3}{\text{kPa}} = P\ (\text{kW h}/1000\ \text{m}^3)$$

The total contacting power requirement, P_T, for a scrubber is, in general, made up of two parts. One is the pressure loss of the gas passing through, P_G, and the other is that of the spray liquid during atomization, P_L, that is, $P_T = P_G + P_L$,

Table 6.1 Scrubber Efficiency Correlation Parameters[a]

Curve in Fig. 6.22	Aerosol	Scrubber	Correlation parameter α	Correlation parameter β
	Lime kiln dust and fume (Kraft mud kiln)			
1	Raw gas (lime dust and soda fume)	Venturi and cyclonic spray	3.49	1.05
2	Prewashed gas (soda fume)	Venturi, pipeline, and cyclonic spray	2.17	1.05
3-A	Talc dust	Venturi	4.00	0.362
3-B		Orifice and pipeline	3.64	0.362
4	Black liquor recovery furnace fume, cold scrubbing water (humid gases)	Venturi and cyclonic spray	2.91	0.620
5	Hot fume solution for scrubbing (humid gases)	Venturi, pipeline, and cyclonic spray	1.50	0.861
6	Hot black liquor for scrubbing (dry gases)	Venturi evaporator	1.06	0.861
7	Phosphoric acid mist	Venturi	2.26	0.647
8	Foundry cupola dust	Venturi	2.25	0.621
9	Open hearth steel furnace	Venturi	2.01	0.569
10	Talc dust	Cyclone	1.99	0.655
11-A	Copper sulfate	Solivore (A) with mechanical spray generator	0.996	1.14
11-B		(B) with hydraulic nozzles	1.34	1.06
12	Ferrosilicon furnace fume	Venturi and cyclonic spray	1.27	0.459
13	Odorous mist	Venturi	1.16	1.41

[a] From Semrau (1960).

where

$$P_G \text{ (kW h/1000 m}^3) = 0.2778 \times \Delta p_G \tag{6.25}$$

where Δp_G is the pressure loss across the unit (kPa), and for nozzles not using compressed air (i.e., hydraulic nozzles),

$$P_L \text{ (kW h/1000 m}^3) = 0.2778 \times \Delta p_L (Q_L/Q_G) \tag{6.26}$$

where Δp_L is the pressure in nozzle (kPa), and Q_L/Q_G is the liquid-to-gas flow rate (dimensionless). For nozzles using compressed air (e.g., pneumatic nozzles), the work of compression of the air per 1000 m³ of gas is P_L.

Semrau expressed the efficiency of a scrubber in terms of a dimensionless transfer unit N_t, defined as

$$N_t = \ln[1/(1 - \eta_0)] \tag{6.27}$$

Semrau also plotted the transfer units against the total contacting power for a number of different scrubbers and dusts. Linear relations were obtained on a log–log plot (Fig. 6.22), which can also be expressed analytically:

$$N_t = \alpha P_T^{\beta} \tag{6.28}$$

where α and β are characteristic parameters for the dust being collected.

Figure 6.23 Comparison of performance characteristics of some wet scrubbers (Calvert, 1977a).

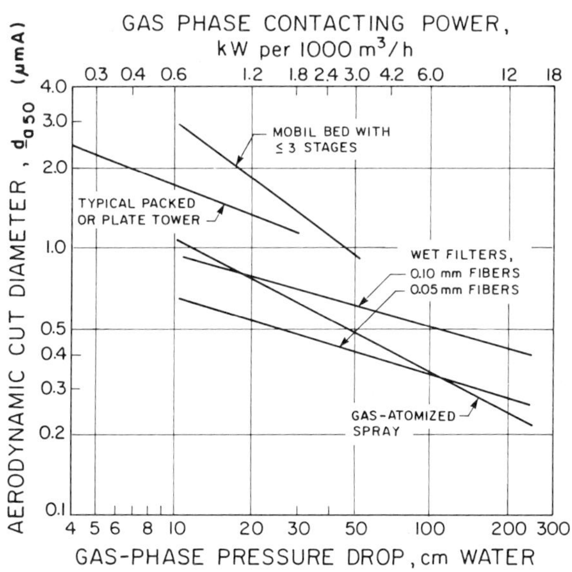

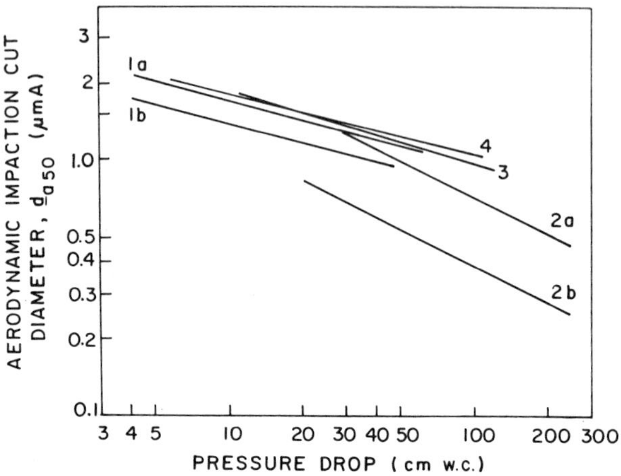

Figure 6.24 Representative cut diameters as a function of pressure drop for several scrubber types. 1a: sieve, $F = 0.4$, $D_h = 0.5$ cm; 1b: sieve, $F = 0.4$, $D_h = 0.3$ cm; 2a: venturi, $f = 0.5$; 2b: venturi, $f = 1.0$; 3: impingement plate; 4: packed column, $D_c = 2.5$ cm (Calvert *et al.*, 1972).

These parameters are listed in Table 6.1. Equation (6.27), with the data listed in Table 6.1, is equivalent to the graph shown in Fig. 6.22.

The contacting power concept has been applied by Calvert (1977a) to a few different types of scrubbers, with the results shown in Fig. 6.23.[3] It is seen here that the venturi-type, gas-atomized spray scrubbers approach an aerodynamic cut diameter of 0.2 μm at a gas phase pressure drop of about 300 H_2O column. It is also seen that a wet filter consisting of a fibrous packed bed of 0.05 mm fibers reaches lower aerodynamic cut diameters (down to about 0.35 μm) than the venturi, for the same gas phase pressure drop. An additional correlation is given in Fig. 6.24, where the aerodynamic impaction cut diameter is plotted against the pressure loss for various scrubbers (Calvert *et al.*, 1972).

6.8 Mist Eliminators

A mist eliminator (entrainment separator, demister) is necessary for every scrubber, because some fine spray, containing dust, is always entrained by the gas. The efficiency of a wet scrubber also depends on how well the mist separator works. Basically, one distinguishes inertial collectors and cyclone collectors. A few of the many available types of inertial mist eliminators are shown in Fig. 6.25. There is an additional pressure loss over the mist

[3] In calculating contacting power, a 45% motor-blower efficiency was allowed for.

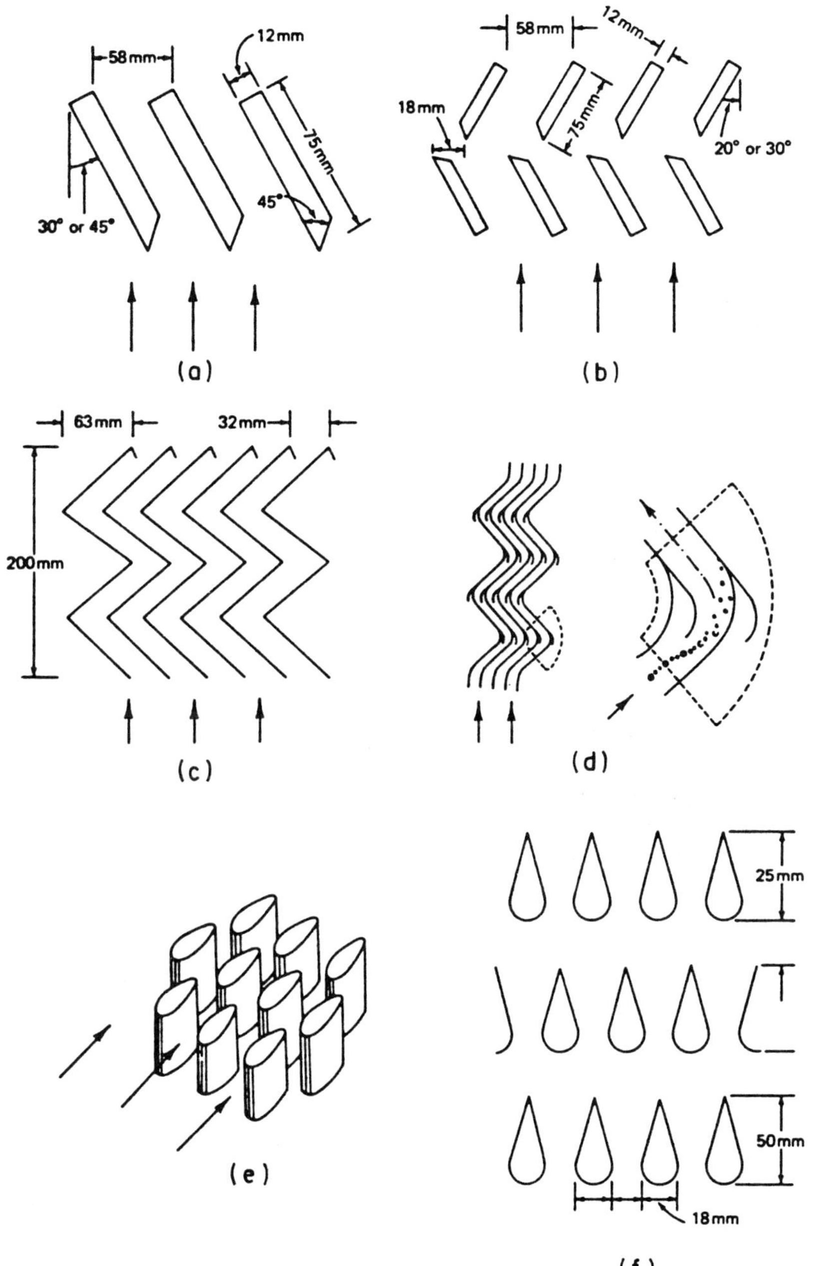

Figure 6.25 Simple inertial separators: (a) single row of baffles, (b) two rows of baffles, (c) W-pattern separator, (d) wave separator (Euroform pattern), (e) vertical rod with teardrop shape in horizontal flow system, (f) horizontal rod with teardrop shape in vertical flow system (Strauss, 1977).

eliminator that must be kept at a minimum. Unfortunately, there is usually a dust buildup in inertial mist eliminators that results in rapidly increasing pressure loss. Dust buildup is combatted by both continuous spraying and periodic cleaning (washing) of the mist eliminator.

6.9 Commercial Aspects

In Table 6.2, a few operating utility scrubbers are listed. There are usually several scrubbing modules in each system. The module capacities are on the order of 100 MW. Originally, most of these scrubbers were installed as particulate scrubbers only, and the SO_2 removal was achieved by pH control. The only flue gas desulfurization (FGD) system in the table is the Cholla station.

Table 6.2 Sample Listing of Operating Wet Scrubber[a]

Utility company[b]	APS	PP & L	PSC	MP & L	APS
Station	Four Corners	Dave Johnston	Cherokee	Clay Boswell	Cholla
Startup date	12/71	4/72	11/72–7/74	5/73	10.73
Application	Particulate	Particulate	Particulate	Particulate	SO_2 + particulate
Design	Venturi	Venturi	3-stage TCA	High-pressure spray	Venturi + packed tower
Scrubber capacity (MW)	575	330	660	350	115
Number of scrubber modules	6	3	9	1	2
Reheated gas	Yes	No	Yes	No	Yes
Liquid-to-gas ratio (liters m^{-3})	1.2	1.7	6.7	1.1	2 to venturi 6 to tower
Δp (cm H_2O)	71	38	25–38	10	51
Electricity required (% of station power)	3–4	2.3	4.0	0.86	2.43
Manpower (total operators)	8				4
SO_2 removal (%)	30	40	20	20	92
Particulate removal (%)	99.2	99.0	95–97.5	99	99

[a] From Gaus (1978) and van Ness *et al.* (1978).
[b] APS, Arizona Public Service; PP & L, Pacific Power and Light; PSC, Public Service Co. of Colorado; MP & L, Minnesota Power and Light.

Wet–dry interface areas in scrubbers are subject to buildup of deposits. Nozzles are installed to spray surfaces that are subject to deposit buildup. Erosion is another problem area. Capital costs of particulate-only wet scrubbers range from \$31 to \$52 per kilowatt in 1977 dollars.

EXAMPLE 6.1

In a spray tower, the air (20°C and 1 atm pressure) flow rate is $Q_G = 29{,}000$ ft^3/min. The liquid flow rate is $Q_L = 360$ imperial gal/min. The tower height is $H = 50$ ft. Assume that the spray water drops have $D = 1$ mm diameter. The dust loading is $L_i = 2.6$ grains/ft^3. Assume a dust particle diameter $d = 10\ \mu$m. The dust density is $\rho_p = 2$ g/cm^3. Estimate

1. the terminal settling velocity of water droplets;
2. the actual falling speed of droplets in the spray tower and the diameter of the tower, D_T;
3. the efficiency of the tower using Fig. 6.3;
4. the collector Reynolds number, Re_c, the interception parameter, R, and the inertial impaction parameter, Ψ;
5. the combined efficiency of inertial impaction and interception; and
6. the efficiency of the tower by Eq. (6.11).

SOLUTION

1. From Table 3.1,

$$u_t = 3.82\ \mathrm{m}^{-1} = 12.5\ \mathrm{ft/s}$$

2. Here,

$$v_L = u_t - v_G$$

v_G is usually in the range 2–5 ft/s. Assume $v_G = 3$ ft/s, then

$$v_L = 12.5 - 3 = 9.5\ \mathrm{ft/s}$$

To calculate D_T, first calculate A_T:

$$A_T = \frac{Q_G}{v_G} = \frac{29{,}000\ \mathrm{ft^3/min}}{3\ \mathrm{ft/s} \times 60\ \mathrm{s/min}} = 161\ \mathrm{ft}^2$$

then

$$D_T = \sqrt{\frac{4}{\pi} A_T} = \sqrt{\frac{4}{\pi} 161} = 14.3\ \mathrm{ft}$$

3. The specific water consumption is

$$W_{sp} = \frac{(360\ \mathrm{gal})/(0.22\ \mathrm{gal/liter})}{(29{,}000\ \mathrm{ft}^3) \times (0.0283\ \mathrm{m^3/ft^3})} = 2\ \frac{\mathrm{liter}}{\mathrm{m}^3}$$

The inlet dust loading is

$$L_i = \frac{(2.6\ \text{grain/ft}^3) \times (0.0648\ \text{g/grain})}{0.0283\ \text{m}^3/\text{ft}^3} = 6\ \frac{\text{gram}}{\text{m}^3}$$

From Fig. 6.3, the exit dust loading is

$$L_e = 1.4\ \text{g/m}^3$$

Thus, the tower efficiency is

$$\eta_0 = \frac{6.0 - 1.4}{6.0} = 76.7\%$$

4. $$\text{Re}_c = \frac{v_\infty \rho D}{\mu} = \frac{u_t \rho D}{\mu} = \frac{(382\ \text{cm/s})(1.2 \times 10^{-3}\ \text{g/cm}^3)(10^{-1}\ \text{cm})}{(1.82 \times 10^{-4}\ \text{g/cm s})} = 252$$

$$R = \frac{d}{D} = \frac{10 \times 10^{-3}\ \text{mm}}{1\ \text{mm}} = 10^{-2}$$

$$\Psi = \frac{\rho_p d^2 v_\infty}{18 \mu D} = \frac{\rho_p d^2 u_t}{18 \mu D} = \frac{(2\ \text{g/cm}^3)(10^{-3})^2\ \text{cm}^2\ (382\ \text{cm/s})}{(18)(1.82 \times 10^{-4}\ \text{g/cm s})(0.1\ \text{cm})}$$

$$\Psi = 2.33$$

5. For spherical collectors,

$$\eta_I = \frac{(2\Psi)^2}{(2\Psi + 0.25)^2} \qquad [\text{high Re}_c\text{; Eq. (4.16)}]$$

$$\eta_I = \frac{(2 \times 2.33)^2}{(2 \times 2.33 + 0.25)^2} = 0.901$$

and

$$\eta_C = (1 + R)^2 - \frac{1}{1 + R} = (1.01)^2 - \frac{1}{1.01} = 0.03$$

$$\eta_{IC} = 1 - (1 - \eta_I)(1 - \eta_C) = 1 - (1 - 0.901)(1 - 0.03) = 0.904$$

(Note that $\eta_{IC} = 0.8$ from Fig. 6.1.)

$$\eta_0 = 1 - \exp\left(-\frac{3}{2}\frac{\eta}{D}\frac{Q_L/A}{u_t - Q_G/A} H\right) \qquad \left(u_t - \frac{Q_G}{A} = u_t - v_G = v_L\right)$$

$$\therefore \eta_0 = 1 - \exp\left[-\frac{3}{2}\frac{0.904}{0.1\ \text{cm}}\frac{(360\ \text{gal/min})(0.16054\ \text{ft}^3/\text{gal})/(161\ \text{ft}^2)}{(9.5\ \text{ft/s})(60\ \text{s/min})} \times (50\ \text{ft})(30.5\ \text{cm})\right]$$

$$\therefore \eta_0 = 0.999998 \approx 1.00$$

Repeat the last calculation for $d = 1\ \mu$m.

$$\therefore\ \Psi = 0.0233$$

$$\therefore\ \eta_I = \frac{(2\Psi)^2}{(2\Psi + 0.25)^2} = \frac{(2 \times 0.0233)^2}{(2 \times 0.0233 + 0.25)^2} = 0.025$$

$$\therefore\ \eta_0 = 0.30$$

When comparing these predicted overall efficiencies with the value based on Fig. 6.3, it must be kept in mind that Fig. 6.3 applies to the hot gases of a blast furnace where there is a great deal of evaporation of spray. Therefore, when applying the value $W_{sp} = 2$ liters/m^3 to Fig. 6.3, only a fraction of this will be left over for dust removal. Equation (6.11), however, assumes that all of W_{sp} is used for collecting dust.

EXAMPLE 6.2

Estimate the specific water consumption, the air velocity in the throat of a venturi, and the throat diameter for minimum pressure loss in the unit. State the total pressure loss and the total contacting power ($f = 0.25$; water pressure to nozzles is 35 kPa).

These data are given: gas flow rate (air at 20°C and 1 atm pressure) = 1.5×10^4 m^3 h^{-1}; dust loading = 50 g m^{-3} ($\rho_p = 2.7 \times 10^3$ kg m^{-3}); and actual dust cut diameter = 0.5 μm.

SOLUTION

Use Fig. 6.19:

$$d_a = d(\rho_p C)^{1/2} = 0.5(2.7 \times 1.33)^{1/2} = 0.947\,\mu\text{m}\ A$$

From Fig. 6.19, the minimum pressure loss is ~ 65 cm w.c.:

$$v_G = 80\ \text{m s}^{-1}; \qquad Q_L/Q_G \approx 1\ \text{liter m}^{-3} = 10^{-3}$$

The throat diameter is obtained from

$$\frac{D_{\text{throat}}^2 \pi}{4} v_G = Q_G$$

$$D_{\text{throat}} = \left(\frac{Q_G \times 4}{v_G \pi}\right)^{1/2} = \left[\frac{1.5 \times 10^4\ \text{m}^3\ \text{h}^{-1} \times 4}{(80\ \text{m s}^{-1})(3600\ \text{s h}^{-1})\,\pi}\right]^{1/2} = 0.258\ \text{m}$$

The contacting power is

$$P_T = P_G + P_L; \qquad P_G = 0.2778 \Delta p_G$$

and 1 cm w.c. is 0.098 kPa, so

$$\Delta p_G = 0.098 \times 65 = 6.37\ \text{kPa}$$

$$P_G = (0.2778)(6.37) = 1.77\ (\text{kW h}/1000\ \text{m}^3\ \text{gas})$$

and

$$P_L = 0.2778\Delta p_L \frac{Q_L}{Q_G} \frac{\text{kW h}}{1000 \text{ m}^3 \text{ gas}} = (0.2778)(35)(10^{-3}) \cong 0$$

$$P_T = 1.77 \frac{\text{kW h}}{1000 \text{ m}^3 \text{ gas}}$$

EXAMPLE 6.3

Estimate the contacting power necessary for a venturi scrubber to remove the particular pollutants from black liquor recovery furnace fumes with 95.3% efficiency.

SOLUTION

Use Fig. 6.22 and Table 6.1. Using curve 4, $\alpha = 2.91$ and $\beta = 0.62$, so

$$N_t = \alpha P_T^{\beta}$$

But,

$$N_t = \ln[1/(1 - \eta_0)] = \ln(1/0.047) = 3.0576$$

Thus,

$$P_t = \left(\frac{N_t}{\alpha}\right)^{1/\beta} = \left(\frac{3.0576}{2.91}\right)^{1/0.62} = 1.083 \frac{\text{kW h}}{1000 \text{ m}^3}$$

or

$$\Delta p_t = 3.9 \text{ kPa} = 39.5 \text{ cm w.c.} = 16 \text{ in. water gauge}$$

EXAMPLE 6.4

1. A fly ash laden gas stream is to be cleaned by a venturi scrubber, using a liquid-to-gas ratio of 8.5 gal/1000 ft^3. The fly ash has a particle density of 0.7 g/cm^3. Use a throat velocity of 272 ft/s and a gas viscosity of 1.5×10^{-5} lb/ft/s. The particle size distribution is given in this table.

Particle size range (μm)	Weight (%)
<0.10	0.01
0.10–0.50	0.21
0.50–1.00	0.78
1.10–5.00	13.00
6.00–10.00	16.00
10.00–15.00	12.00
16.00–20.00	8.00
>20.00	50.00

Determine the average overall collection efficiency, η_T.

HINT: Calculate an average water drop diameter using Eq. (6.21) and $f = 0.25$.

2. Estimate the pressure loss across the unit.

HINT: Use the Bernouilli equation to develop the formula for the pressure loss.

SOLUTION

1. $$D = \frac{1920}{v_G}\left(\frac{\sigma}{\rho_L}\right)^{1/2} + 597\left(\frac{\mu_L}{\sqrt{\rho_L \sigma}}\right)^{0.45}\left(1000\frac{Q_L}{Q_G}\right)^{1.5}$$

where v_G is in ft/s.

$$\frac{Q_L}{Q_G} = \frac{8.5/6.23}{1000} = \frac{1.364\ \text{ft}^3}{1000\ \text{ft}^3} = 1.364 \times 10^{-3}$$

$$D\ (\mu\text{m}) = \frac{1920}{272}\left(\frac{72}{1}\right)^{1/2} + 597\left(\frac{10^{-2}}{\sqrt{1 \times 72}}\right)^{0.45}(1.364)^{1.5}$$

$$= 59.896 + 45.740 = 106\ \mu\text{m}$$

The Cunningham correction factor is significantly different from unity only for particles smaller than about 1 μm. As the proportion of such small particles in the same is very low, we are not committing a significant error by assuming that $C = 1$ throughout.

These equations are used:

$$\eta_0 = 1 - Pt$$

$$Pt = \exp\left[\frac{2Q_L v_G \rho_w D}{55 Q_G \mu_G} F(\Psi, f)\right]$$

$$F(\Psi, f) = \frac{1}{2\Psi}\left[-0.7 + 2\Psi f + 1.4 \ln\left(\frac{2\Psi f + 0.7}{0.7}\right) + \left(\frac{0.49}{0.7 + 2\Psi f}\right)\right]$$

$$\Psi = \frac{v_G d^2 \rho_p}{18\mu D}; \qquad f = 0.25$$

The results are now presented in tabular form.

$\bar{d}$ (μm)	η_0	w (%)	$\eta_0 w$
0.05	1×10^{-5}	0.01	1×10^{-7}
0.30	0.011	0.21	2.31×10^{-3}
0.80	0.262	0.78	2.0436×10^{-1}
3.0	0.941	13.0	1.2233×10
7.5	0.985	16.0	1.576×10
12.5	0.989	12.0	1.1868×10
17.5	0.990	8.0	7.92
>20.0	1.000	50.0	50.00
			$\eta_T = 0.98$

The total efficiency is 98%.
Sample calculation, $\bar{d} = 0.05\ \mu$m:

$$\Psi = \frac{272\ \frac{\text{ft}}{\text{s}}\left(0.05 \times 10^{-6}\ \text{m}\ \frac{1}{0.305}\ \frac{\text{ft}}{\text{m}}\right)^2\left(0.7\ \frac{\text{g}}{\text{cm}^3} \times 62.42\ \frac{\text{lb/ft}^3}{\text{g/cm}^3}\right)}{(18)\left(1.5 \times 10^{-5}\ \frac{\text{lb}}{\text{ft s}}\right)\left(106 \times 10^{-6}\ \text{m}\ \frac{1}{0.305}\ \frac{\text{ft}}{\text{m}}\right)}$$

$$= 3.4 \times 10^{-3}$$

$$F(\Psi, f) = \frac{1}{(2)(3.4 \times 10^{-3})}\left[-0.7 - \frac{(2 \times 3.4 \times 10^{-3})(0.25)}{1.7 \times 10^{-3}} + 1.4 \ln\left(\frac{1.7 \times 10^{-3} + 0.7}{0.7}\right) + \left(\frac{0.49}{0.7 + 1.7 \times 10^{-3}}\right)\right]$$

$$= -5.7 \times 10^{-7}$$

$$Pt = \exp\left[\frac{2 \times 1.364 \times 272\ \text{ft/s} \times 62.43\ \text{lb/ft}^3 \times 106 \times 10^{-6}\ \text{m}\ \frac{1}{0.305}\ \text{ft}}{55 \times 1000 \times (1.5 \times 10^{-5}\ \text{lb/ft s})} \times (-5.7 \times 10^{-7})\right]$$

$$= 0.99999$$

$$\eta_0 = 1 - Pt = 1 \times 10^{-5}$$

2. Consider the situation in the throat of the venturi:

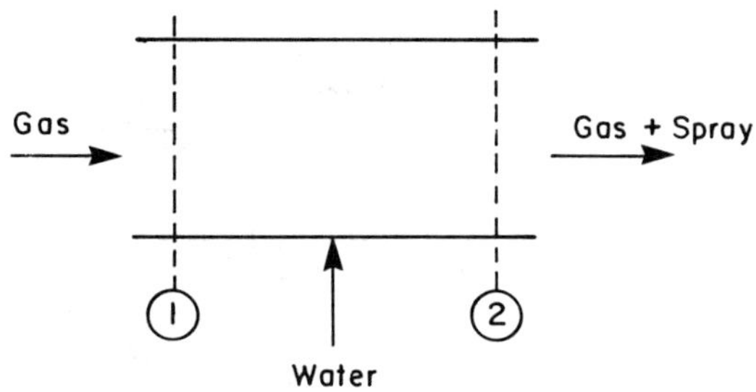

Scheme 6.2

Mechanical energy balance over the throat of the venturi scrubber yields

$$A_t v_1 p_1 + A_t v_1\left(\tfrac{1}{2}\rho_1 v_1^2\right) = A_t v_2 p_2 + A_t v_2\left(\tfrac{1}{2}\rho_2 v_2^2\right)$$

Whence, if it is assumed that $v_1 = v_2 = v_G$,

$$p_1 - p_2 = \tfrac{1}{2} v_G^2 (\rho_2 - \rho_1)$$

But,

$$\rho_1 = \rho_G$$

and

$$\rho_2 = \rho_{MIX} = \frac{\dot{m}_G + \dot{m}_L}{Q_G + Q_L} \approx \frac{1 + (\dot{m}_L/\dot{m}_G)}{Q_G/\dot{m}_G} = \rho_G + \rho_G \frac{\dot{m}_L}{\dot{m}_G}$$

where $\dot{m}_G$ and $\dot{m}_L$ are the mass flows of the gas and the water, respectively. Thus.

$$\Delta p = p_1 - p_2 = \frac{1}{2}\left(\rho_G + \rho_G \frac{\dot{m}_L}{\dot{m}_G} - \rho_G \right) v_G^2 = \frac{1}{2} \frac{Q_L \rho_L}{Q_G} v_G^2$$

If it is assumed that because of water evaporation and other causes the actual pressure loss is double this value, we obtain

$$\Delta p = \frac{Q_L}{Q_G} \rho_L v_G^2 = 1.364 \times 10^{-3} \times 10^3 \ \frac{\text{kg}}{\text{m}^3} \ (272 \times 0.305)^2 \ \frac{\text{m}^2}{\text{s}^2}$$

$$= 9.3875 \times 10^3 \text{ Pa} \approx 38 \text{ in.} \quad H_2O \text{ gauge}$$

The pressure drop formula of Hesketh, Eq. (6.24), predicts for $A_t = 0.1$ m^2,

$$\Delta p = 23 \text{ in.} \quad H_2O \text{ gauge}$$

It has been noted that Eq. (6.23), with a correction factor of 2, overpredicts the actual pressure loss.

PROBLEMS

1. A conventional spray tower is being used to remove a metallurgical fume. This information is available to you:

$$H = 20 \text{ m}$$

$$D_t = 6 \text{ m}$$

$$Q_G = 1.67 \times 10^5 \text{ m}^3 \text{ h}^{-1}$$

$$L/G = 1 \times 10^{-3}$$

$$D = 1000 \text{ m}$$

$$d = 1 \text{ m}$$

$$\mu = 2.5 \times 10^{-4} \text{ g/cm s}$$

$$\rho = 1.2 \times 10^{-3} \text{ g/cm}^3$$

$$\rho_p = 2 \text{ g/cm}^3$$

Estimate

a. the value of the inertial impaction parameter, Ψ;

b. the individual droplet collection efficiency by inertial impaction, η_I;

c. the ratio η_I/D; and

d. the overall collection efficiency of the tower, η_0.

2. It has been suggested that the overall collection efficiency of the tower described in problem 1 can be improved using a finer spray in the tower, for example $D = 100$ m. Evaluate this suggestion by making these calculations.

   a. Estimate the value of Ψ, η_I, and η_I/D.

   b. Can one use the original flow rate? If not, why not? If not, let $v_G = v_L$.

   c. If the flow must be changed, what fraction of the original flow can the tower handle under the new conditions?

   d. How many towers of the original size are now required to handle the total gas flow rate?

   e. What is the diameter of a new tower if only one tower were to be used?

   f. What is the new overall efficiency, η_0, using the new values of η_I/D, Q_L, and Q_G?

3. Design a venturi-type scrubber (i.e., determine the *throat diameter* and the *pressure loss*) for the treatment of 50,000 ACFM (at 68°F) of gas flow. The mean particle diameter is 0.5 μm, the particle density is 187 lb/ft^3, and the required overall removal efficiency is 98%.

   HINT: Use Eqs. (6.19), (6.20), and (6.21) with $f = 0.25$, $\mu_L = 10^{-3}$ Pa s and $\sigma = 72$ mN/m. As this calculation involves trial and error, calculate the throat diameter to two significant places only.

4. The particle size distribution of a fly ash has been measured.

Particle size range (μm)	Weight (%)
< 0.10	0.01
0.10–0.50	0.30
0.50–1.00	0.62
1.00–5.00	18.00
5.00–10.00	25.00
10.00–20.00	39.00
> 20.00	17.00

An overall collection efficiency of 99% must be reached with the help of a venturi scrubber. The particle density of the fly ash is 1.2 g cm^{-3}. The temperature is 80°C and the correlation coefficient k in Eq. (6.22) can be assumed to be equal to 20. Design the venturi while considering the upper limit set on the pressure drop, which is 1000 cm H_2O column.

5. The inlet loading of talc dust in a gas stream is 2 g m^{-3}, which is to be reduced to 0.01 g m^{-3}. Estimate the amount of kW h energy required on an annual basis to do this job for a yearly average gas flow rate equal to 500 m^3 h^{-1}.

   HINT: Assume 45% motor-blower efficiency.

6. Estimate the contacting power necessary for a venturi scrubber to remove particulate pollutants from black liquor recovery furnace fumes with 95.3% efficiency. What difference does it make if hot fume solution is used for scrubbing (recirculated liquor) rather than cold scrubbing water? Would you expect the humidity of the gases being scrubbed to affect the contacting power? If so, can you present an explanation for your answer?

References

Anonymous. (1961). *Chem. Eng. (N.Y.)* **68**(18), 84.

Calvert, S. (1977a). How to choose a particulate scrubber. *Chem. Eng.* **84**(18), 54–68.

Calvert, S. (1977b). Scrubbing. *In* "Air Pollution" (A. C. Stern, ed.), 3rd Ed., Vol. IV, Chap. 6. Academic Press, New York.

Calvert, S., Goldshmid, J., Leith, D., and Mehta, D. (1971). "Scrubber Handbook," NTIS No. PB-213-016. U.S. Depart. of Commerce, NTIS, Springfield, Virginia (USI DE89 72 W271).

Doyle, H., and Brooks, A. F. (1957). The Doyle scrubber. *Ind. Eng. Chem.* **49**, A57–A64.

Dullien, F. A. L. (1977). The role played by eddy diffusion in the aerodynamic capture of particles. *Proc. Particle Technol. Eur. Congr. Nuremberg, 28–30 March, 1977.*

Dullien, F. A. L., and Spink, D. R. (1978). Gas Scrubber and Method of Operation, U.S. Patent 4,067,703.

Gaus, K. S. (1978). Pollution control with SO_2 recovery. *Pollut. Eng.* **10**, 51–55.

Hausberg, G. (1961). Wet dusting in the metallurgical industry. *Staub* **21**, 418–425.

Hesketh, H. E. (1974). Atomization and cloud behavior in wet scrubbers. *U.S.–U.S.S.R. Symp. Control Fine Particulate Emissions, San Francisco, January 15–18.*

Hesketh, H. E. (1979). "Air Pollution Control," p. 227. Ann Arbor Science, Ann Arbor, Michigan.

Jackson, S., and Calvert, S. (1966). Entrained particle collection in packed beds. *AIChE J.* **12**, 1075–1078.

Johnstone, H. F., and Eckman, F. O. (1965). Collection of aerosols in a venturi scrubber. *Ind. Eng. Chem.* **43**, 1358–1363.

Johnstone, H. F., Feild, R. B., and Tassler, M. C. (1954). Gas absorption and aerosol collection in a venturi atomizer. *Ind. Eng. Chem.* **46**, 1601–1608.

Jones, W. P., and Anthony, A. W. (1952). *In* "United States Technical Conference on Air Pollution" (L. C. McCabe, ed.), p. 318. McGraw-Hill, New York.

Marshall, W. R. (1954). Atomization and spray drying. *Chem. Eng. Prog. Monogr. Ser.* **50**.
Nukiyama, S., and Tanasawa, Y. (1938). Experiment on atomization of liquid by means of air stream, I. *Nippon Kikai Gakkai Ronbunshu B Hen* **4**(14), 86–93.
Ray, A. B. (1952). *In* "United States Technical Conference on Air Pollution" (L. C. McCabe, ed.), p. 355. McGraw-Hill, New York.
Schifftner, K. C., and Hesketh, H. E. (1983). "Wet Scrubbers." Ann Arbor Science, Ann Arbor, Michigan.
Semrau, K. T. (1960). Correlation of dust scrubber efficiency. *J. Air Pollut. Control Assoc.* **10**(3), 200–207.
Stairmand, C. J. (1950). *Trans. Inst. Chem. Eng.* **28**, 130.
Strauss, W. (1977). Mist elimination. *In* "Air Pollution" (A. C. Stern, ed.), 3rd Ed., Vol. IV. Academic Press, New York.
Taheri, M., and Calvert, S. (1968). Removal of small particles from air by foam in sieve-plate column. *J. Air Pollut. Control Assoc.* **18**, 240–245.
van Ness, R. P., *et al.* (1978). "Project Manual for Full-Scale Dual-Alkali Demonstration at Louisville Gas and Electric Co.—Preliminary Design and Cost Estimate." EPA-600/7-78-010.

7

Electrostatic Precipitators

7.1 Introduction

Electrostatic precipitators (ESP) are widely used in a large number of industries. They are found in the primary smelter industry and in lead blast furnaces, oil roasters, reverbatory furnaces, and cement kilns. They are also used for the removal of fly ash from coal-fired boilers for electric power generation.

An electrostatic precipitator removes suspended particles from gases by exposing them to a high-voltage electric field. As shown schematically in Fig. 7.1, the basic element of an electrostatic precipitator consists of a concentric tube electrode (or parallel electrode plates) with a discharge electrode wire hanging exactly in the middle. The discharge electrodes usually have a high negative voltage impressed upon them, usually 40,000 V or more, while the collection tubes or plates are grounded. As the particles pass by the discharge electrodes, they become charged and are propelled toward the collecting plates. The rate of migration of the charged particles to the collecting plates depends on the magnitude of the charge on the particle, the strength of the electric field, the size of the particle, and the conditions of gas flow. After the particles are deposited on the collecting electrodes, they lose their charge and the agglomerated dust is removed from the collectors and falls into a hopper.

The electric field is provided by the application of a high dc voltage to the dual electrode system. This is obtained with the help of a transformer–rectifier set. On the discharge electrode (wire), electrical corona discharge, which provides the necessary charges, is established. The particles become charged electrically because of the corona discharge. Once charged, the particles are driven by the action of the electric field toward the grounded electrode or collector.

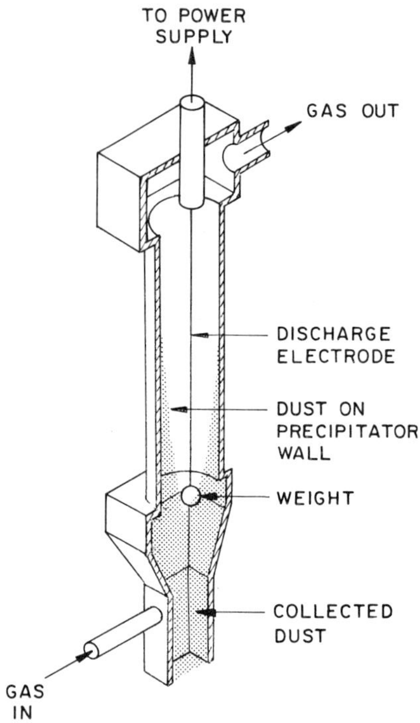

Figure 7.1 Schematic of a wire and pipe precipitator (Ogelsby and Nichols, 1977).

7.2 The Charging of the Particles

A fraction of the particles present in the atmosphere already have electric charges because of contact with naturally occurring, charged gaseous ions and electrons. The average number of charges on a particle increases with the particle size. In a high-intensity electric field, the ion and electron concentrations can become very large, resulting in a greatly increased average number of charges on a particle.

An electrical corona is the discharge associated with the gaseous breakdown in a very strong electric field (e.g., Craggs and Meek, 1953). The corona can be seen in the dark as a luminous glow that takes on a variety of shapes, such as bright spots, brushes, streamers, and uniform glows. One distinguishes negative and positive electrode coronas by the polarity of the corona electrode. A negative corona tends to produce localized tufts, bright spots, or streamers; whereas, a positive electrode corona generally gives a uniform glow over the surface.

The phenomenon of corona discharge depends on the presence of electrons and positive ions in the gas, which are generated by naturally occurring radiation. These electrons and positive ions start moving under the influence of an electric field. If the electric field becomes very strong, that happens near a wire with a small radius of curvature, free electrons will be accelerated to velocities sufficient to strip an electron from the outer shell of a gas molecule on impact, creating a positive ion and another free electron. This new free electron may, in turn, be accelerated to a velocity sufficient to cause further impact ionization. This process is called an avalanche, because it is repeated many times, resulting in the production of large quantities of electrons and positive ions within the corona region.

In the case of a negative corona electrode, free electrons produced by the avalanche process are rapidly carried by the electric field from the wire toward the grounded or positive electrode, whereas the positive ions move toward the wire. Gases such as oxygen, water vapor, and sulfur dioxide capture free electrons and produce negative ions, which also move toward the positive electrode. These electrons and negative ions subsequently charge the particles to be collected.

Almost all industrial precipitators are of the negative corona type because of its favorable voltage–current characteristics under most operating conditions. Positive corona precipitators are used mostly for cleaning air in inhabited space because they produce less ozone. Positive coronas are also used on precipitators that operate at temperatures of 1500°F or more, because at high temperatures, their voltage–current characteristics are more favorable.

Particle charging normally takes place in the region between the corona glow and the collection electrode as a result of negative ions colliding with the particles. The mechanism of charging varies with the particle size. Particles greater than 0.5 μm in diameter are predominantly charged by *bombardment* or *field* charging, where as smaller particles are mostly charged by *diffusion* charging (e.g., White, 1963).

In bombardment charging, the gas ions moving along the lines of force between the discharge electrode and the collector electrode are intercepted by the uncharged particles, collide with them, and deposit their charge. Diffusion charging is a result of impact, which is caused by the thermal movement of the gas ions. There is a limiting value of the charge on the particle beyond which it will not increase any further.

7.3 Particle Migration or Drift

Charged particles migrate because, in the electric field between the discharge and the collector electrode, this electric force is acting on them:

$$G_E = qE \tag{7.1}$$

where q is the particle charge in coulombs (C) and E is the electric field strength in V m^{-1}.

The action of this force is resisted by the aerodynamic drag force, which is represented by Stokes' law. In stagnant air or in air in laminar flow, when these two forces become equal, the charged particle has reached its terminal velocity

$$w = CqE/3\pi d\mu \tag{7.2}$$

where C is the Cunningham correction factor. In this equation, the symbol w has been introduced to distinguish this quantity from the terminal velocity of sedimentation under the action of gravity; u_t; w is usually called the migration velocity of the charged particle. For particles of dielectric constant ε and fully charged by bombardment in a field of strength E, one can write (Lowe and Lucas, 1953)

$$q = \frac{3\varepsilon}{\varepsilon + 2} \frac{d^2}{4} E \tag{7.3}$$

and Eq. (7.2) becomes

$$w = \frac{\varepsilon C E^2 d}{4\pi(\varepsilon + 2)\mu} \tag{7.4}$$

The direction of the velocity of the gas is parallel to the collection electrode, whereas the direction of the migration velocity is perpendicular to that flow. Therefore, the net effect is that the average particle velocity is the result of the two component velocities as shown in Fig. 7.2. In turbulent flow, the trajectory of the individual dust particles follows an irregular path rather than a straight line. The resultant velocity, shown in Fig. 7.2, is valid however, if it is interpreted as representing the average velocity of a very large number of particles, providing that the relaxation time τ of the

Figure 7.2 Velocity components for a charged particle (Ogelsby and Nichols, 1977).

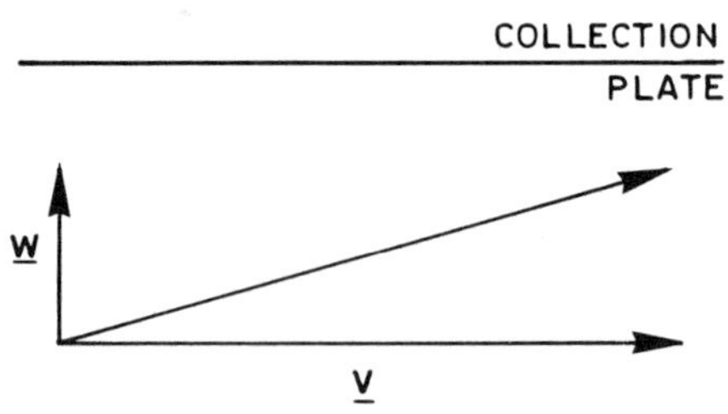

particles is much greater than the time scale of velocity fluctuations. Otherwise, the steady state assumed for w is never established, and as a result, the migration velocity is less than the value predicted by Eq. (7.2) and can even be equal to zero.

EXAMPLE 7.1

Estimate the theoretical migration velocity w of particles 50 μm in diameter with a density of 8×10 kg m^{-3} in still air at 1 atm pressure and 334°C ($\mu = 3.13 \times 10^{-5}$ Pa s) in an electric field of strength $E = 100$ kV m^{-1}, if each particle carries a charge of $q = 1.6 \times 10^{-14}$ C.

SOLUTION

$$w = \frac{CqE}{3\pi d\mu} = \frac{1 \times 1.6 \times 10^{-14}\ \text{C} \times 10^5\ \text{V m}^{-1}}{3\pi \times 50 \times 10^{-6}\ \text{m} \times 3.13 \times 10^{-5}\ \text{Pa s}}$$

$$= 0.1085\ \frac{\text{VC}}{\text{m}^2\ \text{Pa s}} = 0.1085\ \frac{\text{Nm}}{\text{m}^2\ (\text{N/m}^2)\ \text{s}} = 0.1085\ \text{m s}^{-1}$$

7.4 The Design Equation of Electrostatic Precipitators

Theoretically, it should be possible to design pecipitators to give 100% collection efficiency. In reality, however, the particle removal efficiency is an exponential relationship, called the Deutsch equation (Deutsch, 1922), of the form

$$\eta_0 = 1 - \exp[-(A/Q)w'] \tag{7.5a}$$

Here, w' is the effective migration velocity or *precipitation rate parameter*, which is obtained empirically from measured precipitator efficiencies by solving the Deutsch equation for w'. The factor A is the precipitator's collector plate area and Q is the gas flow rate. Consistent units must be used for A, Q, and w'. (Over the years, electrostatic precipitator manufacturers have been collecting extensive data on effective migration velocities, and they have been using these for the design of precipitators.)

Sometimes the Deutsch equation is written in different forms, for example (Hesketh, 1979),

$$\eta_0 = 1 - \exp\left[-\left(\frac{36A}{Q}\right)^n w'\right] \tag{7.5b}$$

where n lies between 0.5 and 1. In this form, A is in m^2, Q in m^3/h, and w' in cm/s.

The original Deutsch equation is based on a steady state differential shell mass balance on the particulates. The shell length in the direction of gas flow is dL, and the shell cross section perpendicular to the direction of the gas flow (for parallel plate precipitators; see Example 7.2) is $2SH$. Then, the (negative) accumulation of particulates dN in the shell is

$$-dn = 2SHv\,dN \tag{7.6}$$

where N is the particulate loading (g/m^3). This is equal, with opposite sign, to the amount of particulates dN deposited or the electrode surface contained in the shell. Denoting the electrode surface in the shell by $2H\,dL$ and the effective migration velocity in the direction perpendicular to the electrode surface by w', one obtains

$$dn = 2Hw'N\,dL \tag{7.7}$$

Combining Eqs. (7.6) and (7.7) and rearranging gives

$$\frac{dN}{N} = -\frac{2Hw'}{2SHv}\,dL \tag{7.8}$$

Integrating between inlet and outlet

$$\int_{N_{\text{in}}}^{N_{\text{out}}} d\ln N = \ln\frac{N_{\text{out}}}{N_{\text{in}}} = -\frac{2Hw'L}{2SHv} \tag{7.9}$$

$$\therefore\ \eta_0 = 1 - \frac{N_{\text{out}}}{N_{\text{in}}} = 1 - \exp[-(A/Q)w'] \tag{7.5a}$$

where $A = 2HL$ and $Q = 2SHv$.

The flow in the bulk of the gases in an electrostatic precipitator is turbulent, causing eddy diffusion of the particulates, which is superimposed on the electric migration. In intensive turbulence, very small particles are always completely mixed, that is, eddy diffusion may completely cancel the effect of electrical migration in the turbulent core. At the same time, these particles may be carried by the eddies to the collector electrode faster than the velocity predicted by Eq. (7.4). It is reasonable to assume that when a particle gets sufficiently close to the collector electrode, where turbulence is greatly diminished, electric migration gradually takes over control of the particle motion and, finally, in the laminar boundary layer, Eq. (7.4) can be expected to apply. Because the particles leave the control volume at the collector electrode surface, only their velocity at the collector surface applies in the Deutsch equation, regardless of what the particle motion is

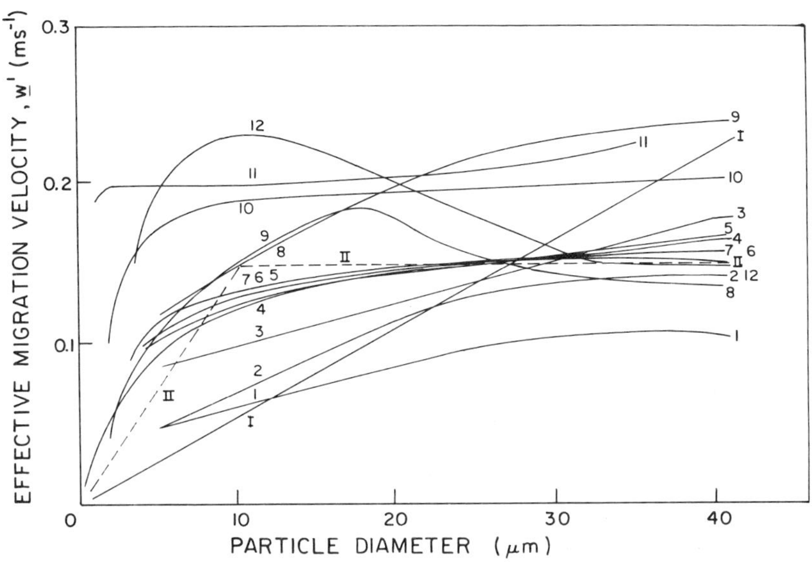

Figure 7.3 Effective migration velocities as a function of particle size for boiler flue dust, catalyst dust, cement dust, and coal dust. I, Theoretical curve based on $E = 130$ kV m^{-1}; II, assumed average curve for average conditions. 1, 2, 3, Boiler flue dust precipitator; 4, catalyst dust precipitator; 5, cement kiln dust precipitator; 6, boiler flue dust precipitator; 7, boiler flue dust precipitator; 8, boiler flue dust precipitator (vertical); 9, 10, boiler flue dust precipitator; 11, coal dust precipitator; 12, boiler flue dust precipitator (vertical) (Heinrich, 1961).

like farther away. In Fig. 7.3, it is interesting to compare the theoretical dependence of w' on the particle diameter (line I) with the actual observed trends, which, on the average, have been represented in the figure by the dashed curve II, consisting of an initial straight line portion of a much greater slope than the theoretical line I, followed by a constant value for particle diameters in the range of 10 to 40 μm. A lack of dependence of w' on particle diameter cannot be explained in terms of electrical migration properties, but it is understandable if it is assumed that the particles are carried directly to the collector electrode surface by eddy diffusion; that is, the particles penetrate the laminar boundary layer deep enough to be in direct contact with the electrode, where their motion is arrested by the mutual attraction between the two opposite charges. It is also understandable, qualitatively speaking, that below a certain size, because of both the smaller inertia and smaller diameter, penetration into the boundary layer may not be deep enough for the particles to reach the electrode surface, so

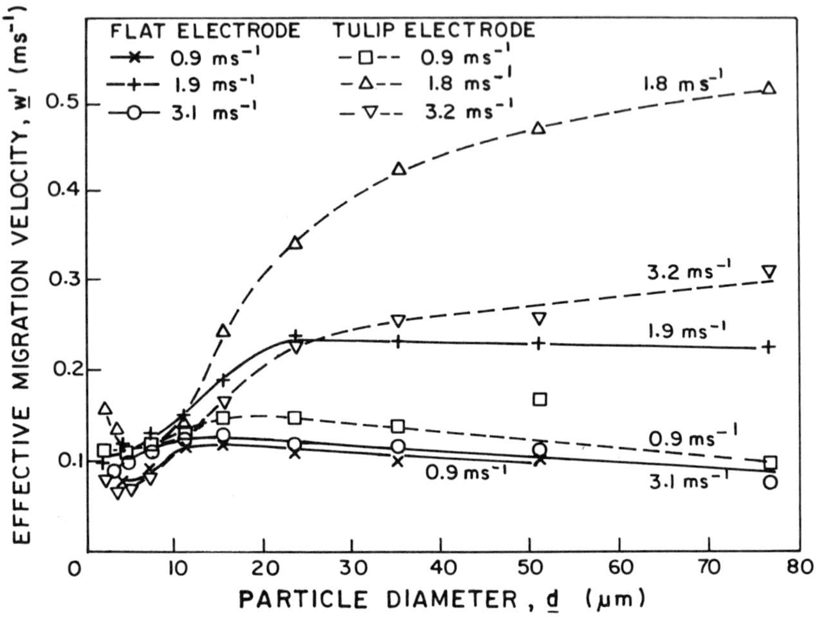

Figure 7.4 Relation between effective migration velocity and particle diameter for different gas velocities with flat and tulip electrodes in a pilot plant (Dalmon and Lowe, 1961).

that electrical migration is necessary for the particles to reach the electrode surface. This behavior may explain the greater rate of increase of w' with particle diameter below a certain particle size than predicted by theory.

Although the theoretical equations indicate that the migration velocity should be independent of the gas velocity, experiments have demonstrated that the latter has, in fact, considerable bearing on the migration velocity as shown, for example, by the curves in Fig. 7.4. It is evident from the diagram that the effective migration velocity has a maximum value at an intermediate gas velocity.

It is recognized (e.g., Hesketh, 1979) that the value of the precipitation rate parameter, w', back-calculated from Eq. (7.5a) as Eq. (7.5b), is affected by a number of factors in addition to particle size, such as gas flow unevenness, dust concentration unevenness within the particle precipitator, charging effectiveness including "sneakage," dust reentrainment, dust resistivity, and any other peculiarities of the specific system.

Gas sneakage is the term used to account for particles that pass through nonelectrified regions of the precipitator. This can occur either at the top or at the bottom of the precipitator. Rapping, used to remove deposited dust from the electrodes, inevitably results in reentrainment of dust. Saltation is

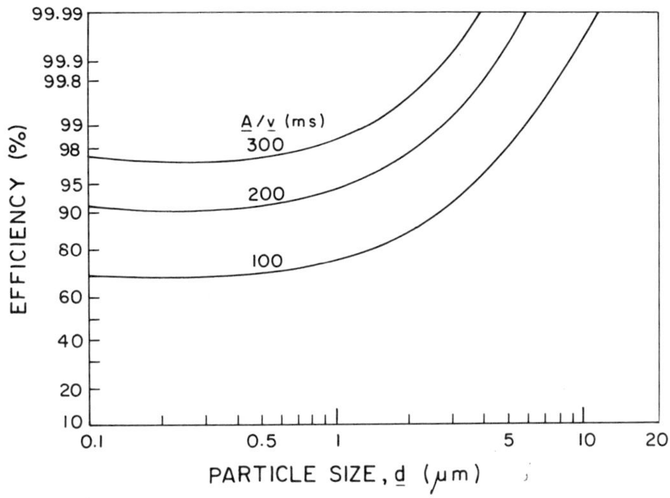

Figure 7.5 Theoretical collection efficiency for various particle-sized fractions (Ogelsby and Nichols, 1977).

the phenomenon that occurs when a particle being deposited strikes and frees already deposited dust, resulting in a decreased precipitation rate parameter.

The effective migration velocity is affected by the electrical field strength, which, in turn, depends on the electrical resistivity of the layer of deposited dust, in addition to the voltage applied. As a result, field strength changes as dust builds up and is removed from both electrodes. The applied voltage can be increased until sparking (spark-over) occurs. Therefore, microcomputers have been introduced to adjust voltages as close to spark-overs as possible, in order to maintain the highest power all the time.

As a result of the variation in migration velocity with the particle size, there is a corresponding change in collection efficiency, that is, larger particles will tend to be collected in the inlet sections of a precipitator. Figure 7.5 shows the theoretical collection efficiency as a function of particle size for a set of typical operating conditions.

EXAMPLE 7.2

One compartment of a plate electrostatic precipitator has five wires per channel spaced at equal intervals. The average drift velocity w' (ft s^{-1}) is given by

$$w' = 0.4(E/10)^2$$

where E is the electric field intensity (kV in.$^{-1}$). The average field strength varies with the distance between charging wires according to

$$E = 10(1 - 0.05L_w)$$

where L_w is the distance (ft) between the charging wires.

The gas velocity is $v = 5$ ft s^{-1}, and the wire-to-plate distance is $S =$ 0.5 ft.

1. Estimate the length of the plates required to obtain an overall efficiency of $\eta_0 = 90\%$.
2. If the last wire in the compartment is broken, what will be the new overall efficiency?
3. The overall efficiency must be maintained at 90%. You suggest that the applied voltage be increased for an interim period until the wire can be reordered and replaced. What new voltage would you recommend?

SOLUTION

1. $$\eta_0 = 1 = \exp\left(-\frac{A}{Q}w'\right) = 0.9$$

$$\frac{Aw'}{Q} = -\ln(0.1) = \ln 10 = 2.303$$

where

$$A = 2LH$$

$$Q = 2SHv$$

$$w' = 0.4(E/10)^2 = 0.4(1 - 0.05L_w)^2$$

where L_w is the spacing between charging wires. Because of the formula given for the variation of the *average* field strength with the distance between charging wires, one must arrange the 5 wires as

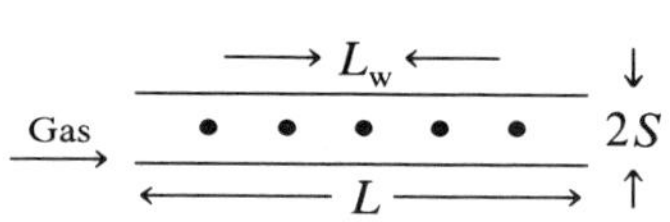

$L = 5L_w$

Therefore,

$$w' = 0.4\left(1 - \frac{0.05}{5}L\right)^2 \frac{\text{ft}}{\text{s}}$$

Combining all these expressions:

$$0.4\left(1 - \frac{0.05}{5}L\right)^2 = 2.303\frac{Sv}{L}$$

$$\therefore 0.4\left(1 - \frac{0.05}{5}L\right)^2 (\text{ft/s}) = \frac{(2.303)0.5\ (\text{ft})\ 5\ (\text{ft/s})}{L\ (\text{ft})}$$

$$\therefore (1 - 0.0100L)^2 = \frac{14.4}{L}$$

Trial and error solution gives

$L = 27$ ft; $\quad L_w = 5.4$ ft

2. The new situation is

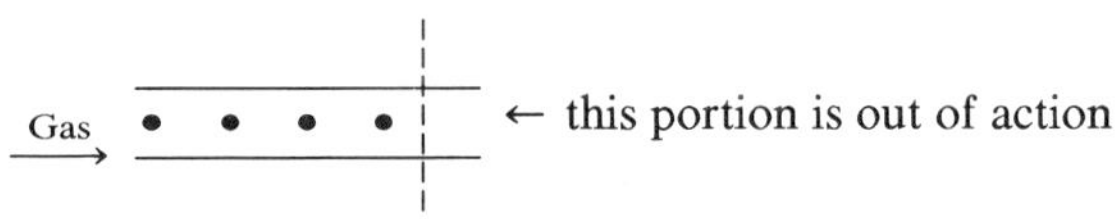

As the same formula for E is to be used, one may assume that the length has been shortened by L_w to $L = 27 - 5.4 = 21.6$ ft.

$$\eta_0 = 1 - \exp\left[-\frac{2LH}{2SHv}0.4(1 - 0.05L_w)^2\right]$$

$$= 1 - \exp\left[-\frac{21.6}{(0.5)(5)}(0.4)(1 - 0.05 \times 5.4)^2\right] = 0.84$$

3. $\ln 10 = \frac{L}{Sv}0.4\left(\frac{E}{10}\right)^2 = \frac{21.6}{(0.5)(5)}0.4\left(\frac{E}{10}\right)^2 = 2.303$

$\therefore E = 8.163$ kV in.$^{-1}$

(compared with the original voltage of 7.301 kV in.$^{-1}$).

7.5 Removal of Collected Particles

The collected particles can be removed from the collection electrode by draining, in the case of liquids, and by irrigating the plates with a liquid or by periodically rapping or vibrating the plates, in the case of solids. Removal of solid particles by flushing the collection plates has been used extensively in some metallurgical processes, especially for blast furnace gases. Wet precipitators have a number of applications, including final mist elimination from gas scrubbers, collection of fine particles from aluminum production, and collection of high-resistivity dust.

The majority of precipitators in use are of the dry type in which dust is removed by rapping. Successful removal of dust by rapping necessitates the formation of a coherent dust, which, when dislodged, falls as an agglomerate. The time between rappings varies from a few minutes to hours, depending on dust loading, dust properties, etc. Rapping results in reentrainment of dry dust.

7.6 Dust Resistivity

Dust resistivity presents some fundamental problems in the operation of a precipitator. Some materials, such as carbon black, that have a very low resistivity will, on contact with a grounded metal surface, lose their charge, and the particles are easily reentrained into the gas stream. Some other materials, such as alumina, have very high resistivity, and these retain their charge for a long time. As a result, the voltage drop across the deposited dust layer can be sufficient to cause breakdown (back ionization) of the interstitial gases within the dust layer, which, in effect, stops precipitator operation. When the resistivity of the deposited dust is in the range of 10^7 to 10^8 Ωm, these problems are avoided.

When certain dusts are very difficult to precipitate, the gases can be conditioned with water, sulfur trioxide, or sulfur dioxide, and other added chemicals (Frisch and Dorchak, 1978), in order to improve the effectiveness of the precipitation. Quite often the gases contain SO_2 and, as a result, also SO_3, which readily dissolves in the water vapor to form a sulfuric acid solution. This will happen because the equilibrium vapor pressure of water over a sulfuric acid solution is much lower than over pure water at the same temperature. Thus, the net effect is condensation of water (to form H_2SO_4 solutions), which is the same as finding the system below the dew point.

Figure 7.6 Elevation of dew point by sulfur trioxide (Müller, 1959).

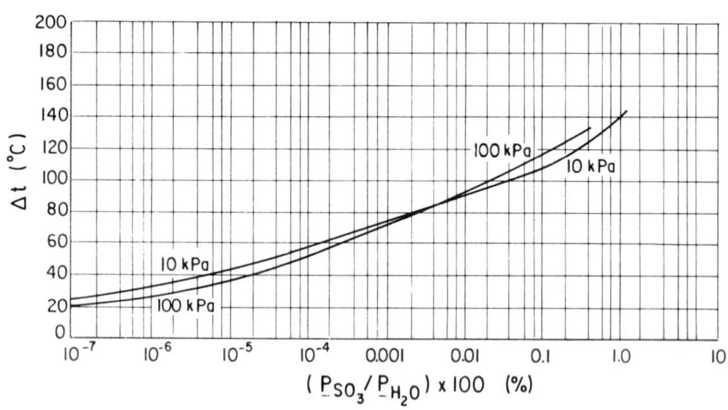

Therefore, the presence of SO_3 results in dew point elevation. The magnitude of the dew point elevation, Δt, caused by the presence of SO_3, has been measured for water vapor pressures of 10 kPa and 100 kPa as a function of the partial pressure ratio p_{SO_3}/p_{H_2O}, and the results are plotted in Fig. 7.6. As it is apparent from these graphs, dew point elevations as great as 100°C may be caused by the presence of SO_3.

7.7 Structural and Design Features and Cost

The main parts of the an electrostatic precipitator are the discharge electrodes, the collecting electrodes, and the enclosing shell. In commercial precipitators, the air flow is either horizontal or vertically upward. The precipitator type, in which the collecting electrodes are vertical tubes (an example is shown in Fig. 7.7), is used either for mist precipitation or,

Figure 7.7 Wet wall electrostatic precipitator with tubular collection electrodes (Ogelsby and Nichols, 1977).

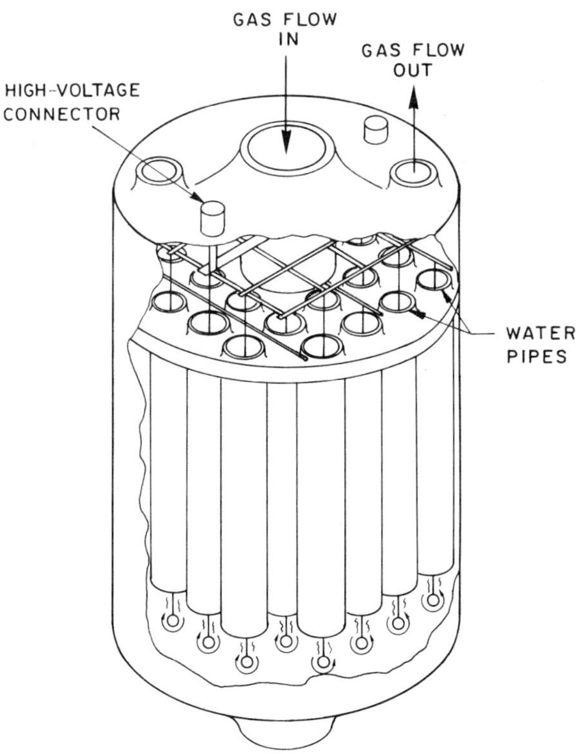

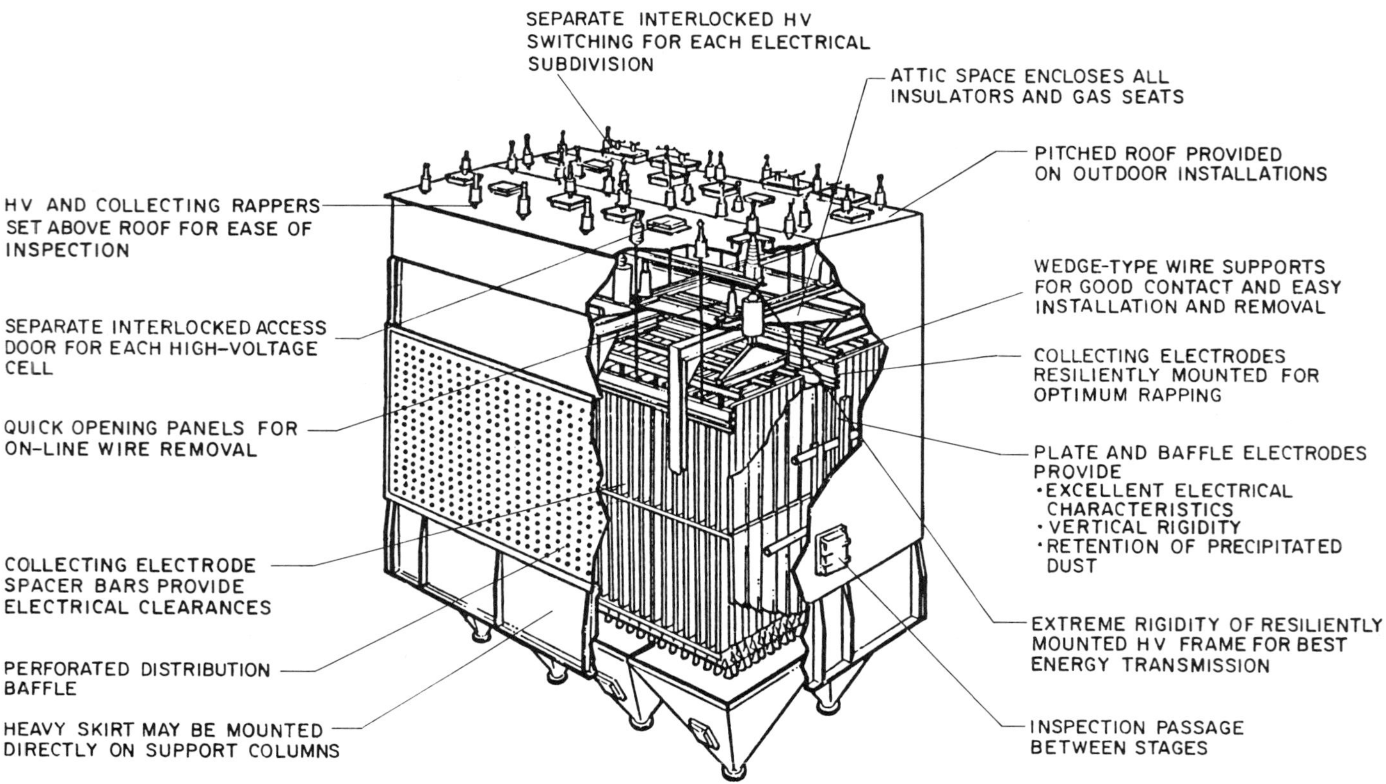

Figure 7.8 Parallel-plate electrostatic precipitator with pyramidal hoppers. (Courtesy of Aerotec Industries, Inc., South Norwalk, Connecticut.)

generally, for relatively small gas flows. For dry dusts, a hexagonal (honeycomb) tube pattern is used, because this pattern permits the most efficient utilization of space. The discharge electrodes in vertical flow precipitators can be fine, round wires; wires with small barbs; or wires with a square or star-shaped cross section held in position by weight hangers on guide frames (Heinrich, 1960). Often, several stages (or fields) are used. For example, in a two-stage vertical tube precipitator, the upper part is the precipitating section, which consists of a heavy wire or a tube without corona formation; attached to the lower end is a fine wire on which the corona forms (Lapple, 1950).

Vertical-flow plate electrodes can be either flat plates or, for very high dust loads, tulip-shaped, (Forrest and Lowe, 1957) in which the dust falls, on rapping, down the space between two plates where there is no gas flow, thereby reducing reentrainment. Horizontal-flow precipitators, which are generally applied for large gas flows, usually consist of various combinations of chambers and fields. The chambers are in parallel, and each chamber is independent from the other chambers, so that gas cannot pass from one chamber to another. Each chamber may contain a number of fields (or stages) in a series, as shown in Fig. 7.8, in order to reduce losses from reentrainment of dry dust freed by rapping. Most horizontal-flow precipitators are of the dry type, but wet electrostatic precipitators, in which the dust is washed off the collecting electrodes with the help of spray nozzles, are also used.

The American and the European precipitator designs are basically different. The American discharge electrodes are usually wires, as seen in Figs. 7.7 and 7.8; whereas, in the European design, as shown in Fig. 7.9, they are tubes, rods, bars with sharp corners or edges, or spikes, and one is in the form of coiled wire. The spiked electrodes improve the operation when the dust resistivity is high. The precipitator shells may be steel, brickwork, or reinforced concrete, depending mainly on temperature, corrosion, and economics.

Precipitator gas velocities vary depending on the ability of the dust to form stable agglomerates and the desired efficiency. Maximum allowable velocities range from 50 cm/s for carbon black, to over 300 cm/s for cement, to 500 cm/s for liquids. These values may be compared with effective migration velocities, ranging from 3 cm/s to 20 cm/s.

For a 23-cm plate-to-plate spacing, operating voltage varies from 25 to 45 kV and from 20 to 35 kV for cold-side and hot-side precipitators, respectively. Hot precipitators operate at temperatures greater than 350°C, whereas cold operation is at 120–150°C.

As shown in Fig. 7.10, the specific collection area, A/Q, in $m^2/A\ m^3$ per second varies from about 20 to 200 for pulverized coal boilers, while the

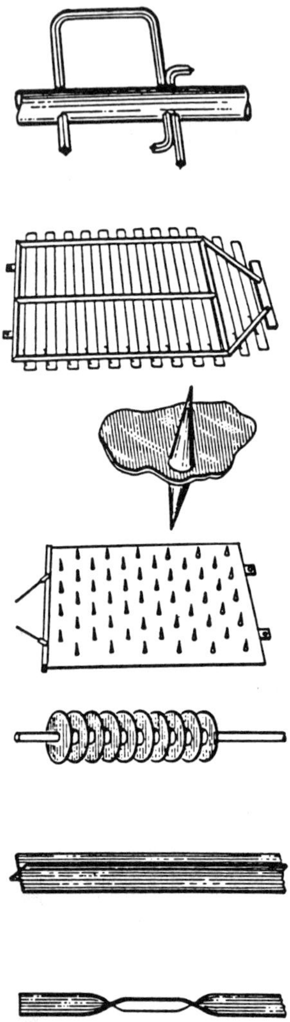

Figure 7.9 Examples of European designed discharge electrodes. [Reprinted with permission from Batel (1972).]

collection efficiencies lie in the range of 95% to 99.9%. Fig. 7.11 shows typical capital cost data of cold-side precipitators for pulverized coal boilers in 1976 dollars. As it is apparent, the cost is about $80/m² of collecting plate area. Efficiencies for a number of industrial precipitators are listed in Table 7.1. The pressure loss in precipitators is very low. It is generally of the order of 10 to 200 Pa.

Primary advantages of electrostatic precipitators are low power requirement (gas pumping and corona discharge), high efficiency, flexibility (can be

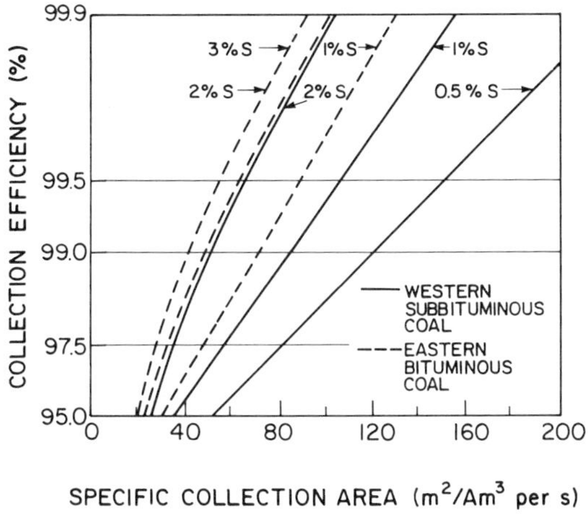

Figure 7.10 Approximate efficiencies of cold-side electrostatic precipitators for pulverized coal boilers (Szabo and Gerstle, 1977).

Figure 7.11 Approximate capital costs of cold-side electrostatic precipitators for pulverized coal boilers in 1976 dollars (Szabo and Gerstle, 1977).

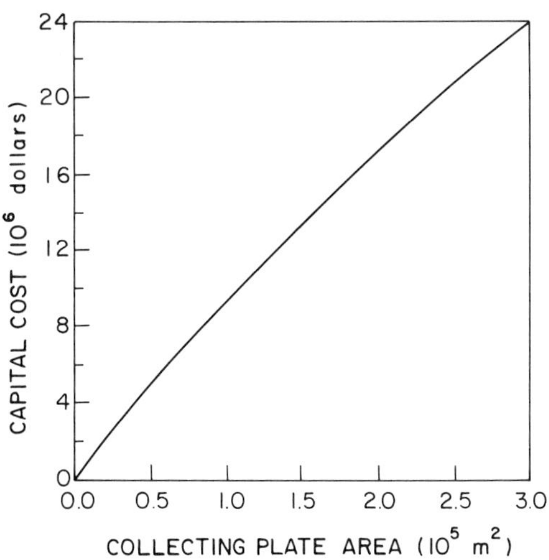

Table 7.1 Typical Performance Data for Electrostatic Precipitators[a]

Type of plant	Gas flow (m^3 h^{-1} at operating temperature)	Dust concentrations (g m^{-3} at operating temperature) Inlet	Outlet	Collecting efficiency (%)	Power consumption (W/1,000 m^3 h^{-1})
		1. *Power stations*			
Pulverized fuel fired boilers	275,000	13.16	0.163	98.67	115.6
Pulverized fuel fired biolers	245,000	10.95	0.062	99.43	131.6
Refuse burning boilers	85,000	16.80	0.58	96.6	140.4
Lignite stoker fired boilers	400,000	1.6–2.0	0.017–0.037	98.15	120.4
Lignite pulverized fuel fired boilers (hammer mills)	1,600,000	4.63	0.16	96.5	40.1
		2. *Coal industry*			
Lignite rotary-type steam dryer	29,000	35.24	0.28	99.25	30.0
Lignite rotary-type steam dryer	26,000	18.31	0.090	99.40	28.1
Lignite plate-type steam dryer	25,000	7.87	0.063	99.20	25.0
Combustion gas lignite dryer	42,000	14.40	0.110	99.50	50.1
Lignite mill dryer	40,000	25.1	0.332	98.67	80.2
Lignite conveying system dedusting	20,500	54.7	0.278	99.40	20.0
Bituminous coal tube type steam dryer	43,000	16.30	0.086	99.50	70.2
Bituminous coal conveying system dedusting	11,000	22.4	0.150	99.30	290.9
Bituminous coal-coke grinding plant	4,800	13.85	0.056	99.59	300
		3. *Coal gas industry*			
Peat gas producer	4,500	5.34	0.008	99.85	702
Cracking plant for natural gas	8,700	0.244	0.002	99.20	120.4
Producer gas from lignite briquettes	13,000	37.7	0.20	99.47	652
Producer gas from semibituminous lignite	48,000	28.7	0.10	99.7	602
Shale-gas cleaning plant	33,980	40.0	0.006	99.9	903
Coke oven town gas cleaning	3,100	24.15	0.010	99.9	903

Table 7.1 Continued

Type of plant	Gas flow ($m^3\ h^{-1}$ at operating temperature)	Dust concentrations ($g\ m^{-3}$ at operating temperature) Inlet	Outlet	Collecting efficiency (%)	Power consumption ($W/1{,}000\ m^3\ h^{-1}$)
Coke oven town gas cleaning	2,300	17.0	0.003	99.9	1,605
Coke oven gas cleaning	14,000	28.0	0.078	99.8	752
Oil carburetted water gas cleaning	12,000	4.73	0.039	99.2	1,404
Tar carburetted water gas cleaning	4,000	10.0	0.050	99.5	1,805
		4. *Paper industry*			
Black liquor burning plant	132,500	2.84	0.13	95.3	265
		5. *Cement industry*			
Rotary kiln dry process, 520 ton/day	136,000 (0.5–1 m/s)	20.7	0.193	99.06	90.3
Lepol rotary kiln dry process, 470 ton/day	127,000 (0.8–1 m/s)	6.33	0.074	98.85	30.0
Rotary kiln wet process, 350 ton/day	145,000 (1.2–1.5 m/s)	21.3	0.067	99.68	85.5
Rotary kiln with calciner, wet process, 350 ton/day	145,000 (0.5–0.9 m/s)	11.0	0.24	98.2	100
Vertical kiln	125,000	1.8	0.048	97.3	90.3
Raw material dryer	36,000	49	0.119	99.75	471
Cement mill	24,000	51.3	0.085	99.8	450.8
Packing machine	15,000	37.7	0.110	99.7	471
		6. *Chemical industry*			
Pyrites roaster, 25 ton/day	9,500	3.22	0.047	98.5	683
Pyrites roaster, 29 ton/day	11,000	2.10	0.037	98.3	802
Pyrites roaster, 35 ton/day	13,000	1.20	0.0036	99.7	962
Pyrites roaster, 36 ton/day	14,000	4.11	0.036	93.5	552
Acid mist from sulfur burning furnace, 7.5 ton/day	2,500	6.88	0.041	99.4	802
Sulfuric acid mist following cooler tower	4,300 (1.2–1.5 m/s)	12.3	0.061	99.5	953
Sulfuric acid mist following cooler tower	5,000 (1.2–1.5 m/s)	7.57	0.070	99.1	729
Blende roaster	15,530	5.13	0.075	98.5	513
Arsenic and sulfuric acid mist removal	14,000	2.71	0.000005	99.99	856

Table 7.1 Continued

Type of plant	Gas flow (m^3 h^{-1} at operating temperature)	Dust concentrations ($g\ m^{-3}$ at operating temperature) Inlet	Outlet	Collecting efficiency (%)	Power consumption ($W/1{,}000\ m^3\ h^{-1}$)
Tail gas for sulfuric acid concentration	23,500	12.37	0.05	99.6	360
Elemental sulfur fume from hydrogen-sulfide combustion plant	4,300	25.8	0.20	99.2	1,700
		7. Mineral earths and salts processing			
Bauxite dryer, 180 ton/day	20,000	15.5	0.05	99.69	115
Bauxite calcining and processing kiln, 220 ton/day	44,000	5.0	0.06	98.8	180
Alumina calciner with multicyclone precleaner, 45 ton/day	14,500	300	0.03	99.9	465
Potassium chloride dryer	29,000	8.0	0.08	99.0	190
Fuller's earth dryer	30,000	4.32	0.02	99.54	210
		8. Nonferrous Metallurgical industry			
Vertical blast furnace, lead ore	10,000	12.0	0.66	99.5	220
Vertical blast furnace, lead ore	16,000	6.35	0.15	97.5	190
Rotary kiln processing, zinc ores	7,000	40	0.44	98.9	200
Rotary kiln processing, zinc ores	12,500	13.2	0.061	99.53	160
Vertical blast furnace, tin ores	9,700	5.0	0.034	99.29	180
Vertical blast furnace, tin ores	3,600	6.85	0.10	98.70	240
Vertical blast furnace, antimony ores	6,200	3.77	0.007	99.8	421
Copper convertors	14,500	4.53	0.136	97.0	70.2
Rotary kiln for nickel bearing iron ores	45,000	27.7	0.065	99.76	270

[a]From Heinrich and Anderson (1957).

operated at high temperatures and in corrosive atmospheres), and dry collection capability. Disadvantages of electrostatic precipitators are their generally high capital and installation costs and their requirement for large areas of land.

PROBLEMS

1. A parallel-plate electrostatic precipitator, consisting of 4 gas passages 4 ft high and 20 ft deep with an 11-in. plate spacing is to treat 41,800 ACFM of air containing 4.3 grains/ft^3 of carbon black. The particle size distribution of the carbon black and the effective drift velocities are listed in this table:

Wt. % < d	d (μm)	w' (cm s^{-1})
5	5	2.2
10	10	5.9
50	20	8.3
90	40	11.2
99.6	80	13.8

   a. Calculate the overall efficiency of the unit.
   b. Calculate the outlet loading and the yearly amount of carbon black emitted into the atmosphere.

2. A horizontal parallel-plate precipitator consists of a single duct 18 m high and 14 m deep with a 20-cm plate-to-plate spacing. A collection efficiency of 96% is obtained with a flow rate of 180 m^3/min.
   a. Estimate the collection efficiency if the flow rate is increased to 200 m^3/min.
   b. Estimate the collection efficiency if the plate spacing is increased to 25 cm.

3. The electrostatic precipitator described in problem 2 is to be used for a dust with this particle size distribution:

Particle size range (μm)	Wt. %
0–1	5
1–5	14
5–10	19
10–20	28
20–30	21
30–50	7
>50	6

Using curve II in Fig. 7.3, estimate the expected overall collection efficiency.

4. Two electrostatic precipitators are used simultaneously, each to handle a gas flow of 500 m^3 s^{-1}. The effective migration velocity has an average value of 0.1 m s^{-1}, and the efficiency is 98%. Estimate the new efficiency if one of the two precipitators were shut down and the remaining unit were to handle the total load.

References

Batel, W. (1972). "Entstaubungstechnik," p. 85. Springer-Verlag, Berlin and New York.

Craggs, J. M., and Meek, J. D. (1953). "Electrical Breakdown in Gases." Oxford Univ. Press (Clarendon), London and New York.

Dalmon, J., and Lowe, H. J. (1961). Colloques Internationaux du Centre Nationale de la Recherche Scientifique, Grenoble, 27th Sept. Editions du Centre Nationale de la Recherche Scientifique, Paris, 1961.

Deutsch, W. (1922). Motion and change of electricity carriers in the cylinder condenser. *Ann. Phys. (Lapzig)* **68**, 355–544.

Forrest, J. S., and Lowe, H. J. (1957). Mechanical engineer's contribution to clean air. *Inst. Mech. Eng.*, 42.

Frisch, N. W., and Dorchak, T. P. (1978). Impact of fuel on precipitator performance. *Pollut. Eng.* **10**, 63–65.

Heinrich, D. O. (1960). Electro-precipitator for Ajax. *Iron Steel Inst.* **33**, 452.

Heinrich, D. O. (1961). Study on electro-precipitator performance in relation to particle size distribution, level of collection efficiency and power input. *Trans. Inst. Chem. Eng.* **39**, 145.

Heinrich, R. F., and Anderson, J. R. (1957). *Chem. Eng. Pract.* **3**, 464.

Hesketh, H. E. (1979). "Air Pollution Control." Ann Arbor Science, Ann Arbor, Michigan.

Lapple, C. E. (1950). *In* "Chemical Engineer's Handbook" (J. H. Perry, ed.), 3rd Ed., p. 1021. McGraw-Hill, New York.

Lowe, H. J., and Lucas, D. H. (1953). Physics of electrostatic precipitation. *Br. J. Appl. Phys. Supp.* **2**, S40–S46.

Müller, P. (1959). Beitrag zur frage des einflusses der schwefelsaeure auf die tauchgas-taupunkttemperatur. *Chem. Ing. Tech.* **31**, 345–351.

Ogelsby, Jr., S., and Nichols, G. B. (1977). Electrostatic precipitation. *In* "Air Pollution" (A. C. Stern, ed.), 3rd Ed., Vol. IV, Chap. 5. Academic Press, New York.

Szabo, M. F., and Gerstle, R. W. (1977). "Operation and Maintenance of Particulate Control Devices on Coal-Fired Utility Boilers." EPA-600/2-77-129.

White, H. J. (1963). "Industrial Electrostatic Precipitation." Addison-Wesley, Reading, Massachusetts.

8

Gaseous Pollutants

8.1 Introduction

Removal of harmful or obnoxious gaseous constituents from industrial gases constitutes an extremely important category of industrial gas cleaning. From the point of view of the amount of pollutant emitted into the atmosphere from stationary sources, sulfur dioxide, SO_2, is among the first of all pollutants. (Another contender is the fly ash emitted by power plants.) According to available data, the quantity of SO_2 emitted in the United States alone is on the order 30 million tons annually (Slack, 1975). Other important harmful or obnoxious gases include nitrogen oxides, carbon monoxide, fluorine and fluoride compounds, chlorine and chlorides, hydrogen sulfide, ammonia, amines, and the pyridine bases. In addition, it is essential in areas of dense or predominantly residential populations to remove the obnoxious odors, essentially organic nitrogen and sulfur compounds produced by processes such as roasting coffee or rendering offal.

There are three major methods of removing gaseous constituents. Gases may be absorbed in a liquid, adsorbed on a solid surface, or changed chemically into a harmless gas. The last usually involves the combustion of the organic material, either directly or with the assistance of catalysts. The basic mechanism in all these is the diffusion of the particular gas either to the surface of an absorbing liquid, adsorbing solid, or catalyst or to the reaction zone of a chemical reaction. Hybrid processes include such methods as dry scrubbing.

8.2 Gas-Absorption Processes

Calculation of the Number of Transfer Units and Tower Height

Gas absorption is carried out by scrubbing the gases in liquid scrubbers, which, in this case, usually consist of spray towers and packed towers. Occasionally venturi scrubbers or other specialty scrubbers are also used. The characteristics of wet scrubbers have been discussed in Chapter 6.

Gas absorption is one of the unit operations in chemical engineering and as such it is treated in detail in textbooks on mass transfer operations (e.g., Treybal, 1980). The major differences between the conventional unit operation of gas absorption and gas absorption as it is used in industrial gas cleaning include the relatively low concentrations and an emphasis on removal, rather than recovery, of the pollutant gases in gas cleaning operations. Often gas absorption is aided by a chemical reaction in the absorbing liquid.

The reader is assumed to be familiar with Raoult's law and Henry's law, which are limiting laws applicable (in nonideal solutions) to the solvent, present in a very high concentration, and the solute, present in a very low concentration, respectively. In the same solution, when Raoult's law applies to the solvent, Henry's law applies to the solute, and vice versa. According to Raoult's law, the vapor pressure p_B of the solvent B is given by

$$p_B = x_B p_B^0 \tag{8.1}$$

where x_B is the mole fraction of the solvent B in the liquid, and p_B^0 is the vapor pressure of pure solvent B at the same temperature and pressure as the solution. In ideal solutions, Raoult's law applies to both the solute and the solvent. According to Henry's law, the vapor pressure p_A of the solute A is given by

$$p_A = c_A H \tag{8.2}$$

where c_A is the concentration (mol/cm^3) of the solute A in the liquid and H is Henry's law constant, which is a function of the temperature. Units of H may be, for example, (atm cm^3/mol). Unfortunately, Henry's law and Raoult's law do not apply when the solute A reacts with the solvent B or with another solute present in the liquid. Therefore, their utility in gas cleaning by absorption processes is very limited. Of general applicability is the equilibrium constant

$$K = y_A/x_A \tag{8.3}$$

where K changes with concentration (in addition to a dependence on temperature and pressure) and y_A is the mole fraction of the solute A in the equilibrium vapor.

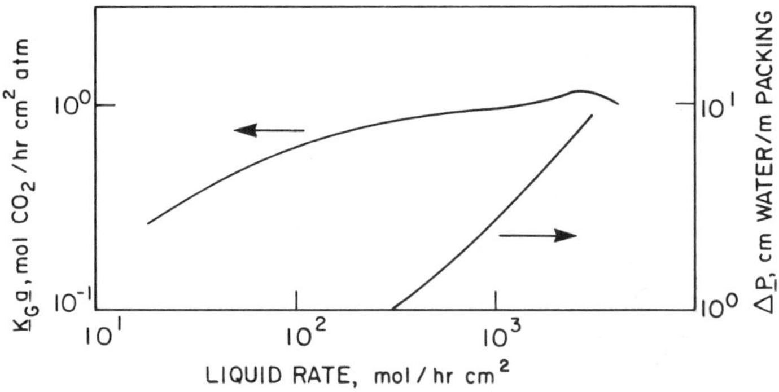

Figure 8.1 K_Ga versus liquid flow rate for CO_2 absorption in 4% NaOH (U.S. Stoneware, 1967).

To reach equilibrium, however, very long times would be required that are not available in a wet scrubber or absorption tower. The amount of pollutant gas removed from the gas and transferred to the liquid in the wet scrubber is less than it would be had equilibrium been reached. In order to calculate the amount of pollutant A transferred, overall mass transfer rate equations can be written as

$$N_A = K_G a \tilde{V}(p_A - p_A^*) \tag{8.4}$$

where N_A is the rate of absorption of A (mol/h), p_A is the partial pressure of A in the gas phase (atm), p_A^* is the partial pressure of A if it were in equilibrium with the prevailing pollutant concentration in the liquid, K_Ga is an empirically determined overall gas-side mass transfer coefficient times the interfacial area a per unit bulk volume of the empty absorber, and $\tilde{V}$ is the total volume of empty absorber. K_Ga versus liquid rate is shown in Fig. 8.1 for CO_2 absorption in 4% NaOH.

The value of N_A varies with different heights in the absorption column, and therefore, it is more convenient to write Eq. (8.4) for a differential section of the column of height dZ and cross section A:

$$d(Vy_A) = K_G ap(y_A - y_A^*)A\,dZ \tag{8.5}$$

where V is molar flow rate of gas, y_A is mole fraction of pollutant A in the gas, p is the total pressure in the gas, and y_A^* is the mole fraction of A in the gas in equilibrium with the liquid solution; $(y_A - y_A^*)$ is represented by the vertical lines BF and AE in Fig. 8.2, where the equilibrium line represents the loci of equilibrium mole fractions of pollutant A in the gas, y_A, and in the liquid, x_A. The operating line represents the actual relationship between y_A and x_A in the column. It can be calculated from given or

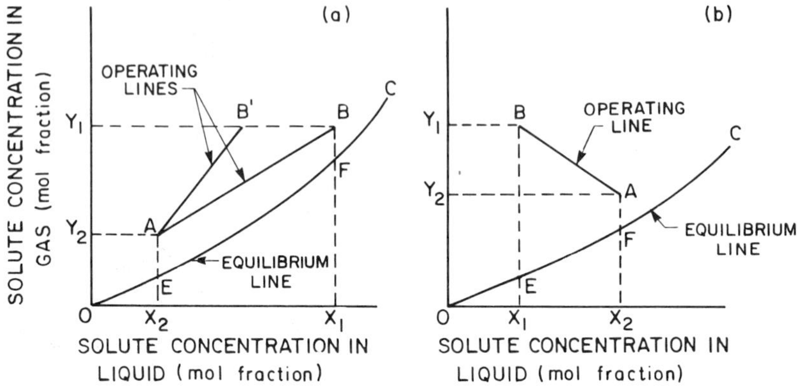

Figure 8.2 (a) Operating and equilibrium lines in a countercurrent column. (b) Operating and equilibrium lines in a cocurrent column. [Reprinted with permission from Strauss (1975).]

assumed compositions and flow rates of the two streams (gas and liquid) at any height of the column by integrating a material balance on pollutant A over the section dZ:

$$d(Vy_A) = d(Lx_A) \tag{8.6}$$

starting from one end, for example, the bottom, of the column:

$$Vy_A - V_1 y_{A1} = Lx_A - L_1 x_{A1} \tag{8.7}$$

Equation (8.7) is valid anywhere between the two ends, 1 and 2, of the column. In general, V and L vary at different heights up the tower, and the operating line is curved. In many gas cleaning problems, however, the pollutant is present in such a small concentration that L and V are practically constant, and the operating line is practically linear. Even though L and V may not be constant, the carrier gas (air) can be considered as practically insoluble in the liquid and, therefore, the molar flow of carrier gas, V_B, can be considered to be constant in the column. Thus,

$$\begin{aligned} d(Vy_A) &= d\left(V\frac{V_A}{V}\right) = dV_A = V_B\, d\left(\frac{V_A}{V_B}\right) = V_B\, d\left(\frac{y_A}{1-y_A}\right) \\ &= V_B \frac{dy_A}{(1-y_A)^2} = \frac{V_B\, dy_A}{(V_B/V)(1-y_A)} = V\frac{dy_A}{1-y_A} \end{aligned} \tag{8.8}$$

Combining this with Eq. (8.5), the following expression is obtained for the height Z of the tower

$$Z = \int_{y_{A2}}^{y_{A1}} \frac{V}{K_G apA} \frac{dy_A}{(1 - y_A)(y_A - y_A^*)} \tag{8.9}$$

It is convenient to divide and multiply under the integral by $(1 - y_A)_{LM}$, where

$$(1 - y_A)_{LM} = \frac{y_A - y_A^*}{\ln(1 - y_A^*/1 - y_A)} \tag{8.10}$$

and writing Eq. (8.9) in the form

$$Z = \frac{V_{av}}{K_G a(1 - y_A)_{LM} pA} \int_{y_{A2}}^{y_{A1}} \frac{(1 - y_A)_{LM}}{(1 - y_A)(y_A - y_A^*)}\, dy_A \tag{8.11}$$

Here, $K_G a(1 - y_A)_{LM}$ is far more nearly constant than $K_G a$ alone. For many situations, $(1 - y_A)_{LM}/(1 - y_A) \approx 1$. Thus, the integral is essentially a measure of the number of times the driving force $(y_A - y_A^*)$ must divide into the change in gas concentration. This is a measure of the difficulty of the mass transfer operation, and the integral is called the number of transfer units, N_t. The group outside the integral is the height of a transfer unit, H_t. The height of the tower is then

$$Z = H_t N_t \tag{8.12}$$

The computation of the number of transfer units for dilute mixtures can be greatly simplified. One proceeds as follows. First of all, the logarithmic mean is replaced with the arithmetic mean, that is,

$$(1 - y_A)_{LM} \approx \frac{(1 - y_A^*) + (1 - y_A)}{2} \tag{8.13}$$

resulting, after integrating by parts, in

$$N_t = \int_{y_{A2}}^{y_{A1}} \frac{dy_A}{y_A - y_A^*} + \frac{1}{2} \ln \frac{1 - y_{A2}}{1 - y_{A1}} \tag{8.14}$$

Next it is realized that the second term in this equation is entirely negligible. Thus,

$$N_t = \int_{y_{A2}}^{y_{A1}} \frac{dy_A}{y_A - y_A^*} \tag{8.15}$$

Now, if the equilibrium curve over the range of compositions can be approximated by a straight line, that is,

$$y_A^* = mx_A + r \tag{8.16}$$

and the operating line may be considered as a straight line as well,

$$y_A = \frac{L}{V}(x_A - x_{A2}) + y_{A2} \tag{8.17}$$

there follows that the driving force $y_A - y_A^*$ is also linear in x_A:

$$y_A - y_A^* = qx_A + s \tag{8.18}$$

Therefore, Eq. (8.15) becomes

$$N_t = \frac{L}{V}\int_{x_{A2}}^{x_{A1}} \frac{dx_A}{qx_A + s} = \frac{L}{Vq}\ln\frac{(y_A - y_A^*)_1}{(y_A - y_A^*)_2} \tag{8.19}$$

giving

$$N_t = \frac{y_{A1} - y_{A2}}{(y_A - y_A^*)_{LM}} \tag{8.20}$$

where $(y_A - y_A^*)_{LM}$ is the logarithmic mean of the concentration differences at the ends of the tower.

EXAMPLE 8.1

A coal gas is to be freed of its light oil (assumed to be benzene) by scrubbing with wash oil as an absorbent. The gas flow rate $Q_G = 30{,}000$ ft^3/h at 800 mm Hg, 75°F, containing 2.0% by volume of benzene. A 95% removal is required. The wash oil enters the scrubber at 75°F, containing 0.005 mole fraction of benzene, and has an average molecular weight of 260. An oil circulation rate of 1.5 times the minimum is to be used. Wash-oil–benzene solutions obey Raoult's law. The scrubber is a packed tower. Determine the number of transfer units N_t. Use the simplified procedure for dilute solutions.

SOLUTION

$$V = 30{,}000\frac{492}{460 + 75}\frac{800}{760}\frac{1}{359} = 80.8\ \frac{\text{mol gas}}{\text{h}}$$

$$y_{A1} = 0.02, \qquad Y_1 = \frac{y_{A1}}{1 - y_{A1}} = \frac{0.02}{1 - 0.02} = 0.0204\ \frac{\text{mol benzene}}{\text{mol dry gas}}$$

$$V_B = 80.8(1 - 0.02) = 79.1\ \frac{\text{mol dry gas}}{\text{h}}$$

For 95% removal of benzene,

$$Y_2 = (0.05)(0.0204) = 0.00102 \text{ mol benzene/mol dry gas}$$

$$x_{A2} = 0.005, \qquad X_2 = 0.005/(1 - 0.005) = 0.00502 \text{ mol benzene/mol oil}$$

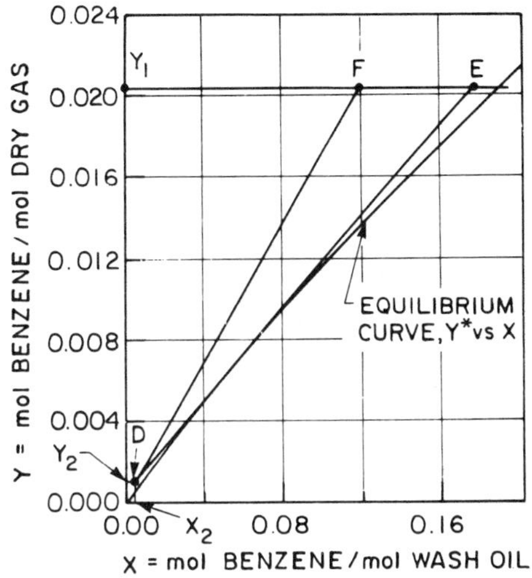

Figure 8.3 Operating lines and equilibrium curve for Example 8.1 (Treybal, 1980).

At 75°F, the vapor pressure of benzene is

$$p_A^0 = 90 \text{ mm Hg} = 90/760 = 0.1184 \text{ atm}$$

Raoult's law is

$$p_A^* = p_A^0 x_A = 0.1184 x_A$$

$$y_A^* = p_A^*/p; \qquad p = 800/760 = 1.053 \text{ atm}$$

$$Y^* = \frac{y_A^*}{1 - y_A^*}; \qquad X = \frac{x_A}{1 - x_A}$$

and

$$y_A^* = \frac{Y^*}{1 + Y^*}; \qquad x_A = \frac{X}{1 + X}$$

Substitution into Raoult's law gives

$$\frac{Y^*}{1 + Y^*} = 0.1250 \frac{X}{1 + X}$$

which is the equilibrium curve for the absorber and is plotted in Fig. 8.3. Operating lines originate at point D in this figure. For the minimum oil rate, line DE is drawn as the line of maximum slope that touches the equilibrium curve (tangent to the curve). At $Y_1 = 0.0204$, $X_1 = 0.176$ mol benzene/mol oil (point E). Hence, the minimum oil circulation rate is

$$\begin{aligned} \min L_B &= V_B(Y_1 - Y_2)/(X_1 - X_2) \\ &= 79.1(0.0204 - 0.00102)/(0.176 - 0.00502) \\ &= 8.95 \text{ mol oil/h} \end{aligned}$$

For 1.5 times the minimum,

$$L_B = (1.5)(8.95)$$
$$= 13.43 \text{ mol oil/h}$$

Hence,

$$X_1 = V_B(Y_1 - Y_2)/L_B + X_2$$
$$= 79.1(0.0204 - 0.00102)/13.43 + 0.00502$$
$$= 0.1192 \text{ mol benzene/mol oil}$$

The actual operating line is shown as DF.

With these data, Eq. (8.20) may now be used to calculate N_t:

$$y_{A1} = 0.02, \qquad x_{A1} = \frac{X_1}{1 + X_1} = \frac{0.1192}{1 + 0.1192} = 0.1065$$

$$y_{A1}^* = \frac{Y_1^*}{1 + Y_1^*} = (0.125)(0.1065) = 0.01331$$

$$y_{A2} = \frac{Y_2}{1 + Y_2} = \frac{0.00102}{1.00102} = 0.00102$$

$$y_{A2}^* = \frac{Y_2^*}{1 + Y_2^*} = (0.125)x_{A2} = (0.125)(0.005) = 0.000625$$

$$(y_A - y_A^*)_2 = 0.00102 - 0.000625 = 0.000395$$

$$(y_A - y_A^*)_{LM} = (0.00669 - 0.000395)/\ln(0.00669/0.000395)$$
$$= 0.00222$$

$$\therefore N_t = \frac{0.02 - 0.00102}{0.00222} = 8.55$$

Cocurrent flow is not customary in packed columns or spray towers. It is the normal mode of operation, however, in venturi or cyclonic scrubbers. The calculation of the number of transfer units in the cocurrent case is similar to that in the countercurrent, except that the operating line has a slope in the opposite direction, as shown by the operating line in Fig. 8.2.

An empirical equation for the number of transfer units in a cyclonic spray scrubber absorbing sulfur dioxide has been obtained by Johnstone and Silox (1947):

$$N_t = \frac{1.04 \times 10^{-5} np}{V^{0.8} A} \tag{8.21}$$

where n is the number of nozzles in the scrubber, A is cross-sectional area of gas inlet (m^2), p is total pressure of gas (kPa), and V is the gas flow rate through entry (kmol s^{-1} m^{-2}).

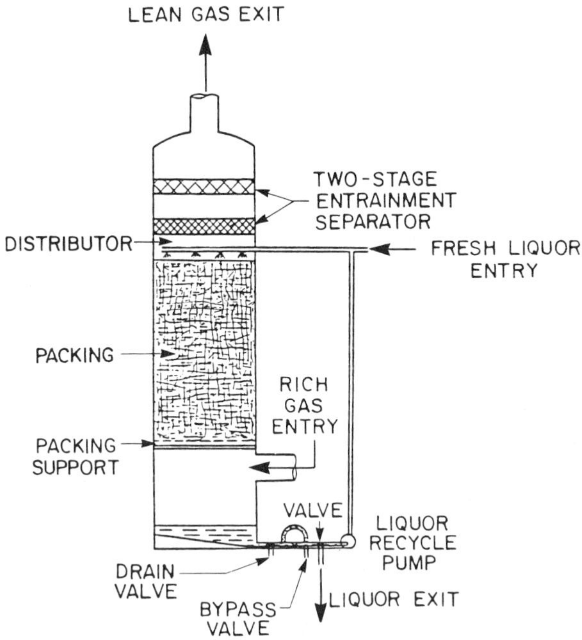

Figure 8.4 Schematic diagram of a packed tower. [Reprinted with permission from Strauss (1975).]

Calculation of Tower Diameter

The diameter of an absorption column depends on two factors: the liquid-to-gas flow rate ratio and economic tower dimensions. In a packed tower, shown schematically in Fig. 8.4, which may be filled with broken stones, rings, saddles, grids, or other types of tower filling packing, the pressure drop suffered by the gas depends on the gas and liquid flow rates in a manner similar to that shown in Fig. 8.5. At moderate gas rates, the curve for each liquid rate is nearly parallel to that for dry packing, but the slope starts to increase at some point and it continues to increase progressively. The point where the slope starts to increase initially is the initial load point, or simply, the loading. At this point, some liquid starts to be held up in the packing, and the liquid holdup keeps increasing with increasing gas flow rate at first slowly and then suddenly more rapidly. At the same time

one of a number of visible changes occurs: (1) a layer of liquid, through which the gas bubbles, may appear at the top of the packing, (2) liquid may fill the tower, starting at the bottom, so that there is a change from the gas-continuous

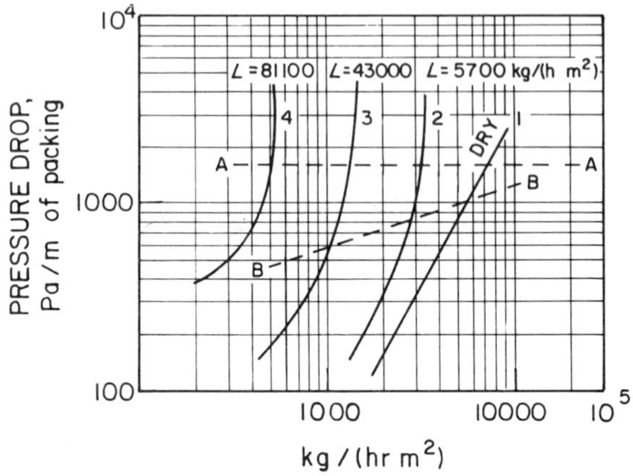

Figure 8.5 Relationship between gas flow rate (V), liquid flow rate (L), and pressure drop (Δp) in a packed tower. [Reprinted with permission from Strauss (1975).] AA is the flooding point line, and BB is the loading point line.

> liquid-dispersed condition to one of liquid-continuous gas-dispersed, or (3) slugs of foam may rise rapidly upward through the packing. At the same time, entrainment of drops of liquid by the gas leaving to the packing at the top increases rapidly, and the tower is said to be "flooded" (Treybal, 1980, p. 194).

It is usual to operate columns at gas flow rates less than 50–60% of the flooding point, and this determines the minimum tower diameter. Flooding rates for a large number of fluids and packings have been correlated in Fig. 8.6 (Eckert, 1970), where

$$X = \frac{L}{V}\frac{M_1}{M_g}\left(\frac{\rho_g}{\rho_1}\right)^{0.5} \tag{8.22}$$

with M_1 and M_g being the molecular weights of the liquid and the gas, respectively. Molar gas flux V' is obtained by

$$V' = \left(\frac{Y\rho_g\rho_1}{M_g^2 F \Psi \mu_1^a}\right)^{0.5} \tag{8.23}$$

where F is the packing factor from Table 8.1, $\Psi = \rho_{H_2O}/\rho_1$, μ_1 is the viscosity of the liquid [g/(cm s)], and $a = 0.2$ when $F \geq 90$ and 0.1 when $F < 90$.

The tower diameter D_T (m) is calculated from the ideal gas-based relation

$$D_T = 1.25\left(\frac{Q}{V}\frac{p}{273 + t}\right)^{0.5} \tag{8.24}$$

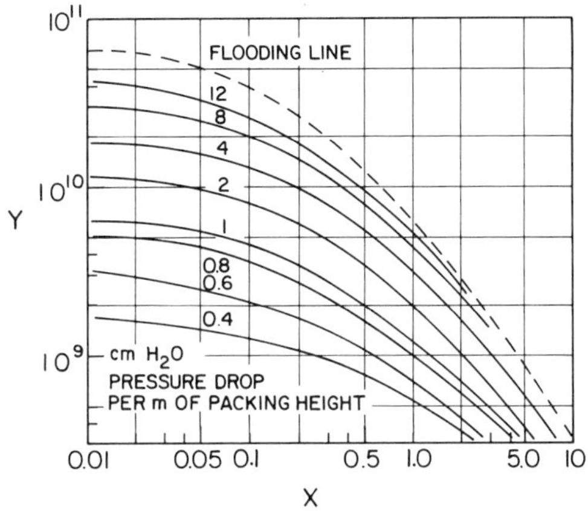

Figure 8.6 Generalized pressure drop correlation for countercurrent packed gas–liquid absorption towers (Eckert, 1970).

with Q being the gas volumetric flow rate (A m^3/h), and t the temperature (°C).

EXAMPLE 8.2

Calculate the tower diameter for a gas (air) flow rate of 3000 A m^3/h at 1 atm pressure and 20°C temperature. A pressure drop of 2.0 cm H_2O per m of packing height is required. Plastic Pall rings, 3.8 cm in size, are to be used. The scrubbing liquid is water.

Table 8.1 Packing Factor, F, for Packed Tower Absorbers for Wet and Dumped Packing[a]

		Nominal Packing Size (cm)							
Packing	Material	0.6	1.0	1.6	1.9	2.5	3.8	5.1	7.6
Intalox saddles	Ceramic	725	330		145	98	52	40	22
Intalox saddles	Plastic					33		21	16
Raschig rings	Ceramic	1600	1000	380	255	155	95	65	37
Raschig rings, 0.08-cm wall	Metal	700	390	170	155	115			
Berl saddles	Ceramic	900			170	110	65	45	
Pall rings	Plastic			97		52	32	25	
Pall rings	Metal			70		48	28	20	
Tellerettes	Plastic							45	

[a]From Eckert (1970).

SOLUTION

Inspection of the problem shows that there is an infinity of solutions. Therefore, the liquid flow rate must also be specified in order to find a unique solution. As it is evident from Example 8.1, the liquid-to-gas flow ratio, together with the tower height, determines the degree of removal of the pollutant gas from the gas stream. Let us suppose that previous calculations of the number of transfer units have resulted in a ratio $L/V = 1.6$ (mol liquid/mol gas).

Hence,

$$X = (1.6)\left(\frac{18}{29}\right)\left(\frac{1.2 \times 10^{-3}}{1}\right)^{0.5} = 0.034$$

Read from Fig. 8.6 at $\Delta p = 2$ cm H_2O per meter of packing height: $Y = 1.03 \times 10^{10}$. Therefore,

$$V' = \left[\frac{(1.03 \times 10^{10})(1.2 \times 10^{-3})(1)}{(29)^2(32)(1)(0.01)^{0.1}}\right]^{0.5} \cong 27 \frac{\text{mol}}{\text{cm}^2\ \text{h}}$$

and the corresponding liquid rate is

$$L' = (1.6)(27) \cong 43\ \text{mol/cm}^2\ \text{h}$$

Finally, the tower diameter D_T is

$$D_T = 1.25\left[\left(\frac{3000}{27}\right)\left(\frac{1}{293}\right)\right]^{0.5} = 0.77\ \text{m}$$

8.3 Absorption Processes for Various Gases

Sulfur Dioxide (SO_2) Removal

Sulfur dioxide is the most important gaseous air pollutant from stationary sources because of the wide occurrence of sulfur in fuels and mineral ores. There are two major sources of sulfur dioxide. In the treatment of sulfur-bearing minerals (lead, zinc, copper, tin, and other ores), sulfur constitutes a major part of the ore, and this is released during sintering and smelting as sulfur dioxide in high concentrations, which causes considerable damage, even at a large distance from the point of release. Restrictions in many parts of the United States stipulate that the stack concentration of sulfur dioxide from a smelter should not exceed 0.75% by volume (Katz and Cole, 1950) and, in compliance with the law, these gases are usually treated.

Much more difficult is the problem of the sulfur dioxide present in the combustion gases from power generation. Here, the concentration of sulfur dioxide is much lower, as shown by the values in Table 8.2. The quantity of sulfur dioxide produced is, however, very large, because of the enormous

Table 8.2 Concentrations of Sulfur Dioxide Combustion Waste Gases[a, b]

Fuel	Sulfur (%)	SO_2 in Flue gas (%)
Coal	1	0.06
Coal	4	0.25
Fuel oil	2	0.12
Fuel oil	5	0.31

[a]Using 15% excess air in combustion.
[b]Reprinted with permission from Strauss (1975).

volumes of stack gas emitted daily. A modern coal-fired power station, with a capacity of 100 MW, burning 9000 tonnes of coal with 2% sulfur per day, produces 360 tonnes of sulfur dioxide each day.

Smelter Gases

The absorption of sulfur dioxide from smelter gases has been satisfactorily tackled in several ways. Unfortunately, sulfur dioxide is only slightly soluble in water, but it is readily soluble in alkaline solutions. Those actually used contain ammonia, xylidine, or dimethyl aniline, although numerous other amines have been suggested (Bienstock *et al*, 1958). The Lurgi "sulfidine process" uses a 1 : 1 xylidine–water mixture as the absorbent and produces pure sulfur dioxide. Equilibrium curves for sulfur dioxide in the xylidine–water mixture and in dimethyl aniline are shown in Fig. 8.7. Flow sheets of the processes are available in Strauss (1975). Ammonia is used in the COMINCO process. The absorption is carried out in a dilute ammonium sulfate solution to which ammonia is added. Sulfur dioxide and ammonium sulfate are produced. The SO_2 is used in the manufacture of sulfuric acid.

Flue Gas Desulfurization (FGD)

By sheer volume, cost, universality, and political impact, flue gas desulfurization (FGD) by wet scrubbing, along with the removal of fly ash, is the most important air pollution control process. It is widely known from media coverage that acid rain is attributed, in part, to SO_2 emissions in flue gases, that is, the gases produced by the combustion of fossil fuel gas, oil, and coal. Ninety-three percent of SO_2 originates from power plants, central heating plants, and home heating (Kaminsky, 1983). The FGD systems have been used since the middle of the 19th century. The number of FGD processes, at different stages of development, is impressive. In addition to wet scrubbing, there are also dry scrubbing, copper oxide, and carbon

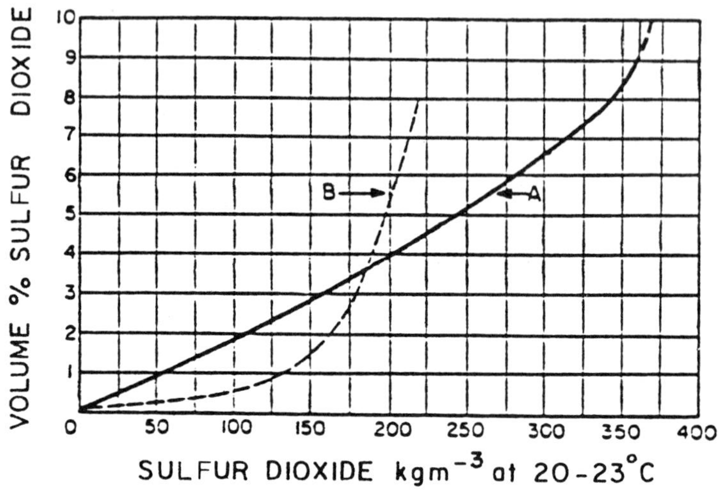

Figure 8.7 Absorption isotherms for sulfur dioxide in (A) anhydrous dimethyl aniline and (B) 1:1 xylidine–water mixture (basis of sulfidine process) (Bienstock *et al.*, 1958).

adsorption processes, just to name a few (Satriana, 1981). A brief description of some of the more important wet scrubbing processes is presented next.

Calcium Sludge or Gypsum Producing Processes. These include all processes that use lime, limestone, or mixtures of calcium hydroxide and magnesium hydroxide. The main reactions are

$$CaCO_3 + SO_2 \rightarrow CaSO_3 + CO_2$$

$$Ca(OH)_2 + SO_2 \rightarrow CaSO_3 + H_2O$$

$$CaSO_3 + \tfrac{1}{2}O_2 + 2H_2O \rightarrow CaSO_4 \cdot 2H_2O \text{ (gypsum)}$$

In addition there is the buffer reaction

$$CaSO_3 + SO_2 + H_2O \rightleftharpoons Ca(HSO_3)_2$$

Over 80% of the operating FGD systems in the United States are of the lime–limestone scrubbing kind. As shown in Tables 8.3a and 8.3b, some produce throwaway sludges, whereas some others produce gypsum. Typical SO_2 removals are about 95%. Data on FGD systems are compiled in Table 8.4.

Another process that has been gaining acceptance in both the United States and Japan (42 major plants in Japan by 1978) are the double alkali processes, also producing either calcium sludge or gypsum. The main

Table 8.3a Summary Description of FGD Processes—Throwaway Category[a]

FGD Process (year of first use)	Primary reagent	Product for disposal	Operating principles	Present status
Limestone/sludge (1933)	$CaCO_3$	$CaSO_3/CaSO_4$ wet sludge	Flue gases scrubbed with limestone slurry. SO_2 reacts to form insoluble calcium compounds, which are separated and partly dewatered for disposal	Commercial operation (coal and oil), principally in United States
Lime/sludge (1972)	CaO	$CaSO_3/CaSO_4$ wet sludge	Flue gases scrubbed with slurry of hydrated lime. SO_2 reacts to form insoluble calcium compounds, which are separated and partly dewatered for disposal	Commercial operation (coal and oil), principally in United States
Double alkali/sludge (1974)	NaOH or Na_2CO_3 and CaO or $CaCO_3$	$CaSO_3/CaSO_4$ wet sludge	Flue gases scrubbed with alkaline sodium based compounds in solution. Used solution regenerated with lime or limestone in separate reactor to form insoluble calcium compounds, which are separated and partly dewatered for disposal	Recent commercial operation on industrial sized coal-fired boilers in United States
Seawater scrubbing (1975)	Seawater	Dilute sulfate solution	Flue gases scrubbed with seawater on a once-through system, diluted, oxidized and neutralized, and discharged back to sea	Commercial operation on small oil-fired boilers in Norway

[a] Reprinted with permission from Satriana (1981).

Table 8.3b Summary Description of FGD Processes—Gypsum Category

FGD process (year of first use)	Primary reagent	Product for disposal or sale	Operating principles	Present status
Limestone/gypsum (1973)	$CaCO_3$	Gypsum	Flue gases scrubbed with limestone slurry. Resulting calcium–sulfur compounds are air-oxidized to gypsum, separated, and dewatered	Commercial operation (coal and oil), principally in Japan
Lime/gypsum (1964)	CaO	Gypsum	Flue gases scrubbed with slurry of hydrated lime. Resulting calcium–sulfur compounds are air-oxidized to gypsum, separated, and dewatered	Commercial operation (coal and oil), principally in Japan; also recently in West Germany
Double alkali/gypsum (1973)	NaOH or Na_2CO_3 and CaO or $CaCO_3$	Gypsum	Flue gases scrubbed with alkaline sodium compounds in solution. Used solution regenerated with lime or limestone in separate reactor, and the insoluble calcium–sulfur compounds are air-oxidized to gypsum, separated, and dewatered	Commercial operation on oil-fired boilers in Japan
Dilute sulfuric acid/gypsum (Chiyoda) (1972)	H_2SO_4 and $CaCO_3$	Gypsum	Flue gases are scrubbed with dilute sulfuric acid containing ferric ion catalyst. Air oxidation produces more sulfuric acid, which is bled off and neutralized with limestone. The gypsum is separated and dewatered	Commercial operation on oil-fired boilers in Japan

[a]Reprinted with permission from Satriana (1981).

Table 8.4 Summarized Data on FGD Systems Installed on Boiler Plants[a]

FGD process	Number of systems surveyed	MWe Range[b]	Predominant fuel		Reported SO_2 removal efficiency range[c] (%)	Reported plant availability range[c] (%)	Process capacity range, (Nm^3/s)[d]
			Type	Sulfur (%)			
Limestone/sludge	3	115–820	Coal	0.5–5.4	57–92	88–94	82–108
Lime/sludge	6	65–835	Coal	0.8–4.7	70–92	83–100	28–139
Double alkali/sludge	2	20–32	Coal	0.9–5.0	90–99	80	25–35
Seawater scrubbing	1	(50)[e]	Oil	3.0	90–97[f]	96–99	17
Limestone/gypsum	6	250–850	Coal	0.6–2.2	90–96	97–100	103–250
Lime/gypsum	2[g]	37–750	[h]	2.9	80–92	96–98	35–333
Double alkali/gypsum	4	26–450	Oil	2.0–2.9	90–95	95–98	23–355
Dilute sulfuric acid/gypsum	2[g]	20–350	[h]	3.6[i]	96	100	270
Sodium sulfite	2	115–200	[h]	2.5–3.5	90–92	97–99	99–172
Magnesium oxide	4	95–155	[h]	2.0–3.8	90–95	[j]	102
Carbon adsorption	2	20–49	Coal	0.9–5.0	70–80	NA[k]	10–42
Copper oxide	1	36	Oil	2.5	80–90	NA[k]	30

[a]Reprinted with permission from Satriana (1981).
[b]Range of MWe units served by FGD system.
[c]Whole numbers.
[d]Range of individual scrubber (or equivalent absorber) capacity.
[e]Equivalent MWe.
[f]Efficiency dependent on bypass (reheat) flow.
[g]Includes one coal-fired demonstration unit.
[h]Equal number of coal and oil firing systems were surveyed.
[i]Calculated value.
[j]Insufficient data base.
[k]NA, Not available.

difference between these FGD processes and the conventional lime–limestone processes lies in the separation of sludge or gypsum. The scrubbing step utilizes an aqueous solution of a soluble alkali, such as Na_2SO_3. The absorption reaction is

$$Na_2SO_3 + SO_2 + H_2O \rightarrow 2NaHSO_3$$

Here, NaOH is formed in the regeneration step. Na_2CO_3 is added as sodium makeup. Both react with SO_2 as shown here:

$$2NaOH + SO_2 \rightarrow Na_2SO_3 + H_2O$$

$$Na_2CO_3 + SO_2 \rightarrow Na_2SO_3 + CO_2$$

Limestone regeneration produces calcium sludge or gypsum. Separating the use of lime from the absorption step reduces scaling and plugging there (see Satriana, 1981). Use of magnesium oxide offers a greater absorption capacity and the possibility of regenerating the MgO by thermal decomposition of $MgSO_3 \cdot 3H_2O$. This process, however, is significantly more expensive than those using lime or limestone.

FGD scrubbing systems often consist of a venturi scrubber followed by an absorption tower or, sometimes, of spray or packed tower absorbers alone.

Empirical relationships have been obtained for the fractional efficiency of SO_2 removal under certain specific operating conditions. For example, for MgO scrubbing (Koehler and Dobe, 1974) in a venturi scrubber at a specific water consumption equal to 5.4 liters/m^3, the SO_2 removal efficiency has been given by

$$\eta = 1 - \exp[(6.86\Delta p^{-1.014} y^{-3.75+0.271 \ln y} 10^{6-0.031\,\mathrm{pH}}) - 3] \qquad (8.25)$$

where Δp is in cm H_2O, y is the inlet SO_2 concentration in ppm, and pH is the pH. The control of pH plays a very essential role in these scrubbing processes.

This expression is for CaO scrubbing of SO_2 in a venturi scrubber (Gleason, 1971):

$$\eta = 0.307 + 0.0457S + 3.75 \times 10^{-3}\Delta p + 0.048W_{sp} + 0.1516I - 0.0258I^2 - 5.98 \times 10^{-3}C_s \qquad (8.26)$$

where S is mol CaO/mol SO_2 input, Δp is in cm H_2O, W_{sp} is the specific water consumption, in liters water/m^3 gas, I is the molarity of Na^+ present, and C_s is the slurry solids concentration, wt. %. The two expressions are that different that they do not permit a meaningful comparison of the efficiency of the two competing processes.

A very serious problem presented by the limestone–lime processes consists of the difficulties, both technical and commercial, encountered when trying to dispose of the products of the scrubbing process. The throwaway products consist of a sludge, whereas the gypsum category poses the problem of finding markets for the gypsum. The sludges are difficult to dewater. The sulfate content of the sludge can be increased by oxidation, resulting in less difficulty with dewatering. Unless fixed by chemical additives, the dewatered sludge reverts, when wetted, to a thixotropic state or, in everyday language, a goo. Considering the amount of SO_2 to be removed from the gases, dumping of the gooey sludge is prohibitive from an environmental point of view.

Wellmann–Lord Process. This process uses Na_2SO_3 solution in an absorption tower, after cooling of the flue gases with water in a venturi. The main reaction is

$$SO_2 + Na_2SO_3 + H_2O \rightarrow 2NaHSO_3$$

Thermal decomposition of the product produces SO_2-rich gas from which high-purity (99.9%) elemental sulfur or sulfuric acid is produced. In 1980, 31 commercial installations were either in operation or under construction in the United States and Japan. For a detailed discussion of this process, see Satriana (1981). A number of other FGD processes that have been developed to a certain stage are listed in Table 8.5, indicating the situation in this field which existed before 1980. Some of these processes are explained here.

Sodium Carbonate Process. The hot flue gases are contacted with an aqueous Na_2CO_3 solution in a spray dryer, where the solution is evaporated to dryness by the hot flue gases. The dry mixture of sodium sulfite, sodium sulfate, and unreacted sodium carbonate is passed through cyclones and an electrostatic precipitator. Then it is fed into the sulfate reducer, after mixing it with crushed coal, which contains a pool of molten sodium salts at about 1800°F. Molten sodium sulfide is the product. The melt is quenched and dissolved by recycled liquor. In a number of additional steps, the sodium carbonate is regenerated with H_2S as a by-product, which is converted to elemental sulfur as the end product. More details of the process can be found in Satriana (1981).

Dry Systems. Dry FGD is defined (Satriana, 1981) as any process that involves contacting a sulfur-containing flue gas with an alkaline material and results in a dry waste product for disposal. This includes (1) systems that use spray dryers for a contactor with subsequent baghouse or electrostatic precipitator collection of the waste products; (2) systems that involve dry injection of alkaline material into contact with the flue gas and

Table 8.5 Major Emerging FGD Processes[a]

Process	Developer	Current level of development	Previous operating experience	Remarks
Aqueous carbonate	Rockwell International	100-MW system (planned)	Mohave pilot plant test program	Full-scale application not yet demonstrated
Catalytic oxidation	Monsanto	100-MW system (terminated)	Wood River test program	No further process development
Chiyoda Thoroughbred 101	Chiyoda International	20-MW system (terminated)	Scholz prototype test program	Development of the process has ceased in favor of a new design concept. (Thoroughbred 121 which employs limestone reagent in a jet bubbler reactor.)
Copper oxide absorption	Shell/Universal Oil Products	Pilot plant	Big Bend test program	Process available for prototype or demonstration application. No systems planned at present
Dual alkali	A. D. Little/Combustion Equipment Associates	277-MW system (under construction)	Scholz prototype test program	Full-scale application not yet demonstrated
	FMC	250-MW system (under construction)	Industrial systems and utility pilot plants	Full-scale application not yet demonstrated
	Buell/Envirotech	575-MW system (under construction)	Gadsby pilot plant test program	Full-scale application not yet demonstrated
Magnesium oxide	Chemico	150-MW system (terminated) 95-MW system (terminated)	Mystic test program; Dickerson test program	Process demonstrated on full-scale oil- and coal-fired boilers. System now offered for commercial application

	United Engineers	600-MW system (planned)	Eddystone test program	Eddystone 120-MW prototype test program still in progress. A full-scale 600-MW system planned for TVA's Johnsonville Station
Sodium carbonate	A. D. Little/Combustion Equipment Associates	Three 115-MW systems operational	Reid Gardner Station	Three full-scale sodium carbonate (trona) FGD systems have been in service on coal-fired boilers at Reid Gardner Station (Nevada Power). System performance has been good
	Universal Oil Products	509-MW system (planned)	Jim Bridger pilot test program	Full-scale sodium carbonate (30% sodium carbonate purge solution from soda ash plant) FGD system planned for Jim Bridger Station
Wellman Lord	Davy Powergas	115-MW system (operational) 375-MW system (operational) 340-MW system (operational)	Crane test program	115-MW NIPSCO/EPA test program in progress at Mitchell. Two full-scale systems have recently started operations at Public Service of New Mexico's San Juan Station
Dry adsorption	Foster Wheeler-Bergbau Forschung	20-MW system (terminated)	Scholz prototype test program	No further development of system reported; further evaluation of the sulfur reduction component (RESOX) in Germany

continued

Table 8.5 Continued

Process	Developer	Current level of development	Previous operating experience	Remarks
Dry collection	Wheelabrator-Frye Rockwell International	410-MW system (planned)	Leland Olds and Bowen Engineering pilot plant test program	Successful testing has resulted in planning of full-scale spray dryer and fabric filter FGD system at the Coyote Station
	Joy/Niro	455-MW system (planned)	Leland Olds pilot plant test program	Successful pilot plant testing has resulted in planning of full-scale system at Antelope Valley. This system will use lime slurry as the reagent in 2-stage atomizer fabric filter design configuration
	Babcock & Wilcox	550-MW system (planned)	Basin electric pilot plant test program	Successful pilot plant testing has resulted in planning full-scale system at Laramie River. This system will use lime slurry as reagent in a configuration which employs an ESP as the second stage collector
	Carborundum/Delaval	Pilot plant	Leland Olds pilot plant test program	Process similar in design to Wheelabrator-Frye/R.I. and Joy/Niro processes. No full-scale applications have been announced

[a] Reprinted with permission from Satriana (1981).

subsequent baghouse or electrostatic precipitator collection; and (3) other varied dry systems that include addition of alkaline material to a fuel prior to combustion or contacting flue gas with a fixed bed of alkaline material.

Dry scrubbing promises to have certain advantages over the wet systems. The product is a dry, solid waste that does not require sludge handling systems. Nevertheless, the environmental impact of the dry waste products can be expected to be similar to that of the lime–limestone wet systems. Dry systems require less equipment than wet systems. Energy requirements of dry systems should be less because of savings in reheating and pumping requirements. A disadvantage consists of a higher cost of reagents because of the requirement of a higher stoichiometric ratio of sorbent to entering SO_2 and also because the reagents employed in dry systems (soda ash, lime, sodium carbonates, and bicarbonates) are significantly more expensive than limestone. However, the minimum equipment and operating requirements for dry injection systems make the process economically attractive.

Copper Oxide Process. In 1980 only one oil-fired boiler in the world (in Japan) was operating with a full-scale copper oxide FGD system, and a 90% SO_2 removal efficiency was reported.

The copper oxide, CuO, is present in a porous alumina support. SO_2 and O_2 react with CuO to form $CuSO_4$. Then the $CuSO_4$ is regenerated to elementary copper with reducing agents such as H_2, CO, and light hydrocarbons. The copper is oxidized by oxygen gas back to CuO. The end product SO_2 is produced in liquid form or is made into elemental sulfur. A potential advantage of this process is its ability for simultaneous removal of nitrogen oxides, NO_x. See Satriana (1981) for more details on this process.

Carbon Adsorption Process. This process is used in Germany, the United States, and Japan under the name of the Bergbau–Forschung process. The reactions, using activated coke, are

Adsorption in fluidized bed

$$\left.\begin{aligned}(SO_2)\text{gas} &\rightarrow (SO_2)\text{ads}\\ \tfrac{1}{2}(O_2)\text{gas} &\rightarrow (O_2)\text{ads}\end{aligned}\right\} + H_2O \rightarrow (H_2SO_4)\text{ads}$$

Regeneration by desorption after mixing the active coke with hot sand at 600–700°C

$$2(H_2SO_4)_{\text{ads}} + C \rightarrow 2SO_2 + CO_2 + 2H_2O$$

The desorption gases contain up to 30 vol % of SO_2. The coke–sand mixture is separated on a vibrating sieve. The end product can be elemental sulfur, H_2SO_4, or liquid SO_2.

A modification of this process consists of adding NH_3 to the flue gases before entering the absorber, resulting in reduction of NO_x and improved SO_2 removal. More details on this system can be found in Satriana (1981).

Removal of NO_x from Flue Gases

Nitrogen oxides are presumed to contribute to acid rain through the formation of nitric acid and ozone in the atmosphere. According to statistics (Schrod *et al.*, 1981), the main contributors to NO_x formation are as follows:

Vehicles	55%
Power plants	28%
Chemical industry	2.6%

The remainder comes from natural sources. Nitrogen oxide emissions can be controlled by primary measures consisting of lower flame temperatures and lower residence times in the combustion zone, and by secondary measures consisting of reduction of NO_x to N_2 or oxidation to nitrite–nitrate. The secondary processes are reviewed here very briefly.

Wet Processes

Since NO is insoluble in water, it is either necessary to use NO-complexing agents or to oxidize NO to NO_2.

Absorption / Reduction. Using iron (II)-ethylene-diamine tetracetate as a complexing agent, simultaneous absorption of SO_2 and NO_x is achieved. The dissolved NO_x is reduced by the sulfite ions formed (Schrod *et al.*, 1981). This process is used in Japan and Germany.

Gas-Phase Oxidation. Gas-phase oxidation of NO to NO_2 can be achieved using, for example, the Walthar process for simultaneous removal of SO_2 and NO_x (Schrod *et al.*, 1981). These reactions take place:

First

$$SO_2 + NH_3 + H_2O \rightarrow NH_4HSO_3$$
$$NH_4HSO_3 + NH_3 \rightarrow (NH_4)_2SO_3$$

Then

$$NO + O_3 \rightarrow NO_2 + O_2$$
$$2NO_2 + 2NH_3 + H_2O \rightarrow NH_4NO_2 + NH_4NO_3$$

In a third scrubber, finally

$$2NO_2 + 4(NH_4)_2SO_3 \rightarrow N_2 + 4(NH_4)_2SO_4$$
$$O_3 + (NH_4)_2SO_3 \rightarrow O_2 + (NH_4)_2SO_4$$

The end product is ammonium sulfate fertilizer.

Dry Processes

Selective Removal of NO_x. The following reactions take place with the help of SCR (selective catalytic reduction) catalyts in the temperature range between 180°C and 450°C:

$$6NO + 4NH_3 \rightarrow 5N_2 + 6H_2O$$

$$6NO + 8NH_3 \rightarrow 7N_2 + 12H_2O$$

Processes exist that use TiO_2-based, iron-oxide-based, and zeolite SCR catalysts (Schrod *et al.*, 1981).

Simultaneous Removal of NO_x and SO_2. Most notable is the Bergbau–Forschung process, using activated coke as the adsorbent (Schrod *et al.*, 1981. The process has two main stages:

1. In the presence of activated coke (adsorbent):

$$SO_2 + \tfrac{1}{2}O_2 + H_2O \rightarrow H_2SO_4 \text{ (adsorbed)}$$

2. In the presence of activated coke + catalyst:

$$6NO + 4NH_3 \rightarrow 5N_2 + 6H_2O$$

Removal of Fluorine and Fluoride Compounds

Hydrofluoric acid gas (HF), silicon tetrafluoride (SiF_4), and the combination of these, fluosilicic acid (H_2SiF_6), are found in the waste gases of superphosphate fertilizer manufacture when the phosphate rock is treated with sulfuric acid; in the smelting of aluminium when fluorspar (CaF_2) is used as a flux; in the electrolytic manufacture of aluminium, where aluminium oxide (alumina Al_2O_3) is fused in cryolite (Na_3AlF_6) at 900–1000°C and electrolyzed; and in calcium metaphosphate furnaces. Fortunately [see the equilibrium diagrams in Fig. 8.8(a) and (b)], both hydrogen fluoride and fluosilicic acid (with which silicon tetrafluoride is in equilibrium) are very readily soluble in water, and water scrubbing is nearly always effective in reducing the fluoride concentration to an acceptable value (Whynes, 1956). Countercurrent spray towers, cocurrent flow spray towers, venturi scrubbers, wet cell washers, and packed towers have all been used in actual plants or pilot plants for treatment of these gases. Correlations of scrubber efficiency with power input are shown in Fig. 8.9.

When fluorine gas is present, the use of water should be avoided because the fluorine does not always react with water, and also because explosions

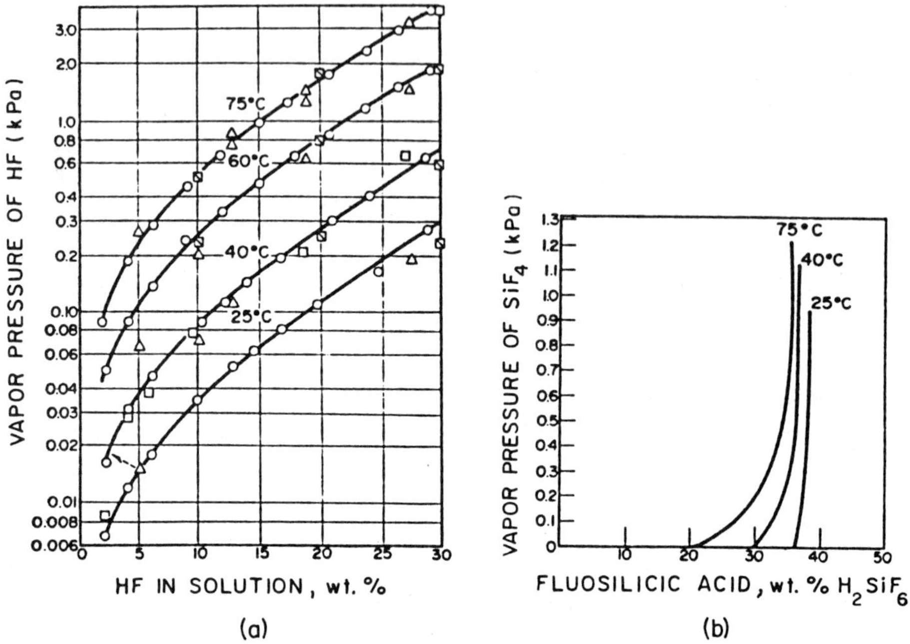

Figure 8.8 Equilibrium diagrams for aqueous hydrofluoric and fluosilicic acid solutions: (a) partial pressure of hydrogen fluoride over hydrofluoric acid solutions; (b) partial pressure of silicon tetrafluoride over aqueous solutions of fluosilicic acid (H_2SiF_6) (Whynes, 1956).

Figure 8.9 Correlation of scrubber efficiency with respect to hydrogen fluoride with power input (Lunde, 1958). Data on silicon tetrafluoride originate from Kohl and Riesenfeld (1960).

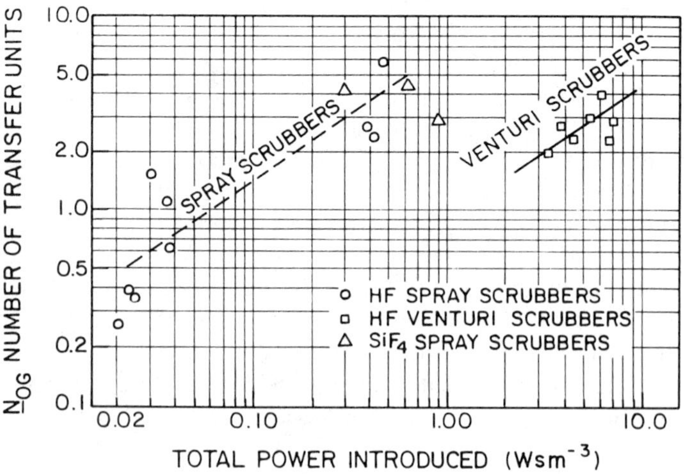

have occurred in some systems. In this case, sodium hydroxide in 5–10% concentrations has been found to be a satisfactory absorbent.

Removal of Chlorine and Chlorides

Absorption of hydrochloric acid gas in water, in which it is very soluble, is used in the manufacture of hydrochloric acid. Chlorides are emitted with the waste gases of smelting aluminium, from scrap where salt (NaCl) is used as flux, and other processes. The absorption of elemental chlorine from this gas is necessary before further processing is possible.

The absorption of hydrochloric acid vapor requires cooling because of greater amount of heat generated in the absorption process. The process is usually a two-stage one. A cooler-absorber, frequently of carbon, is used in the first stage. In the second stage, the tail gases are passed through a packed tower with stoneware packing for the removal of the last traces of the acid (Strauss, 1975).

Removal of Hydrogen Sulfide

Hydrogen sulfide is a highly poisonous gas. Natural gas, refinery gas, and coal gas, all of which are used for industrial and domestic heating as well as chemical processing, contain hydrogen sulfide as their major sulfur impurity. They may also contain smaller concentrations of carbon disulfide (CS_2), carbonyl sulfide (COS), thiophene (C_4H_4S) and thiols (mercaptans, RSH), pyridine bases, hydrogen cyanide, carbon dioxide, and ammonia. Hydrogen sulfide also occurs in the waste gases from the evaporation of kraft pulping liquors and some smelting operations. Gases containing hydrogen sulfide are corrosive when cooled below their dew point, have an objectionable odor, cause difficulties in steel making and heat treatment, and create other problems. It is therefore necessary to remove hydrogen sulfide and some of the other compounds from these gases. Some municipalities require hydrogen sulfide concentrations as low as 0.0115 g m^{-3} for domestic gas, although concentrations of 0.35–0.70 g m^{-3} are generally permissible (Strauss, 1975).

The most common process for purifying (or sweetening, as this is sometimes called) natural and refinery gases, which contain small quantities of carbon disulfide and carbonyl sulfide and no ammonia, is the ethanolamine process (Strauss, 1975). Essentially, hydrogen sulfide and carbon dioxide, both of which are present, are absorbed in a bubble column in an aqueous solution of ethanolamine, forming a complex at low temperatures, which is then passed to a stripping column where, on heating, the acid gases are regenerated and the ethanolamine solution is returned to the absorption column. The first compound to be used was triethanolamine (TEA), but it was subsequently found that for the absorption of both carbon dioxide and hydrogen sulfide, a 15–20% aqueous solution of monoethanolamine (MEA) was more suitable as it has a higher capacity per unit weight of solvent, a higher reactivity, and is more easily regenerated. The H_2S released from the ethanolamine solution is reacted with SO_2 in a Claus reaction to form sulfur and water. The SO_2 is normally generated by oxidation of $\frac{1}{3}$ of the H_2S. The Claus reaction is $2H_2S + SO_2 \rightarrow 3S + 2H_2O$.

In an number of processes, aqueous sodium carbonate solutions are used to absorb H_2S by the reaction

$$Na_2CO_3 + H_2S \rightarrow NaHCO_3 + NaHS$$

The H_2S is stripped from the absorbing liquor at a reduced pressure, using steam.

In the Thylox process (Rees, 1953), sodium thioarsenate is used as both the absorbing and oxidizing agents:

$$H_2S + Na_4As_2S_5O_2 \rightarrow Na_4As_2S_6O + H_2O \qquad \text{Absorption}$$

$$2Na_4As_2S_6O + O_2 \rightarrow 2Na_4As_2S_5O_2 + 2S \qquad \text{Regeneration}$$

In the Stretford process (Nicklin and Holland, 1963), the hydrogen sulfide is absorbed by an alkaline solution containing sodium carbonate, sodium vanadate, and anthraquinone-2,6 and 2,7-disulfonate (ADA).

Removal of Nitrogen Compounds (Excluding Nitrogen Oxides)

Nitrogen compounds in the carbonization of coal are, in part (50–80%), turned into basic nitrogen compounds that are present in the gas. A typical distribution of these compounds in raw town gas is 1.1% ammonia, 0.1–0.25% hydrogen cyanide, and 0.004% pyridine bases, with traces of nitric oxide and some free nitrogen (approximately 1%) also present (Strauss, 1975). As are the sulfur compounds that occur in the same gas in similar quantities, the nitrogen compounds are both corrosive and toxic. Thus, a number of processes have been developed for the simultaneous removal of hydrogen sulfide and ammonia, which produce ammonium sulfate and elemental sulfur.

Ammonia is highly soluble in water and in strong acids, as it reacts with these to give ammonium ions. Because of the reaction, the system does not obey Henry's law, and experimentally determined absorption equilibrium data in the literature should be used. The solubility of ammonia in water, however, falls off rapidly with temperature, and so two-stage recovery systems operating at different temperatures are commonly used.

8.4 Gas-Adsorption Processes

Adsorption consists of the collection of gases and vapors on the surface of solid adsorbents. Since the mechanism of adsorption of gas molecules on the surface of a solid is very complex and depends on the physical and chemical nature of the gas and the solid in each particular case, it is very difficult to deduce a general approach to the design of adsorption equipment and countercurrent adsorption plants. In practice, most of the designs are based either on experience with other similar plants or on pilot plants. Nonetheless, an understanding of the principles is of considerable assistance in deciding whether adsorption is the best process for the removal of certain gases, in the selection of suitable adsorbing materials, and in scaling up designs.

When a gas molecule is adsorbed on the surface of a solid, it is held on the surface either by physically attractive forces (London–van der Waals forces; physical adsorption) or by chemical forces (chemisorption). There is a continuous transition of the types of adsorption between these two limiting types. Physical adsorption is reversible, in the sense that decreasing the partial pressure of the adsorbed gas (adsorbate) from p_2 to p_1 isothermally results in desorption of the same amount of adsorbate as was adsorbed when raising the partial pressure from p_1 to p_2 isothermally. Chemisorption, however, is irreversible in the same sense of the word; that is, a decrease in partial pressure will not result in desorption. Adsorption is an exothermal process, accompanied by the heat of adsorption. The heat of chemisorption is on the order of heats of reaction, that is, it is much greater than the heat of physical adsorption, which is on the order of heats of vaporization of a liquid.

Adsorbents

The solids best suited to adsorption are very porous, with very large effective surface areas, which are obtained with materials such as carbon, alumina, or silica gel. Other desirable characteristics of the adsorbing solids are that the granules should be hard so that they do not collapse under their own weight when packed into a tower, and that they should not powder or fracture easily, so that they can be transported and poured into containers without breakage. The three solids mentioned have these properties. It is frequently a requirement that the solid, after it has been saturated with the gas molecules, can be easily regenerated and reused.

The adsorbents can be classified into three groups:

1. Nonpolar solids, where the adsorption is mainly physical
2. Polar solids, where the adsorption is chemical, without changing the chemical structure of the molecules and the surface
3. Chemical adsorbing surfaces, which adsorb the molecules and then release them after reaction, which may be either catalytic, leaving the surface unchanged, or noncatalytic, with the surface atoms requiring their replacement

Nonpolar Adsorbents—Carbon

The only important material in the group of nonpolar solids is carbon, which consists wholly of neutral atoms of a single species and presents a surface that is a generally homogeneous distribution of electrical charges at the molecular level, without potential gradients. Such a surface does not bind polar molecules in preference to nonpolar ones at particular sites. Carbon is therefore very effective in adsorbing nonpolar organic molecules, particularly near their normal boiling pints. Even if there is water vapor in the gas stream, the organic molecules will be adsorbed preferentially, because the polar water molecules are more strongly attracted to one another than to the nonpolar carbon surface. Thus, large organic molecules are very readily adsorbed, smaller organic molecules and large inorganic ones less easily, smaller inorganic ones still less, and those which are permanent gases hardly at all.

Activated carbon (frequently called charcoal, if the source is wood) is made by the pyrogenic decomposition of suitable coals and woods in special retorts. The raw materials may be lignites and bituminous coals, woods, and the shells of nuts. Some of the purest carbon is made from Eucalyptus Marginata (Australian Jarrah), which is exceptionally low in mineral content, while good quality commercial activated charcoals with very large effective surface areas come from coconut shell, which has been charred at about 1150°C and subsequently activated by treatment with steam at about 600°C.

The basic parameters that determine the action of a carbon are the internal pore surface area and the diameters of the pores, which determine whether the molecules to be adsorbed can actually reach all of the pore surface. The surface area is determined most accurately by the nitrogen gas adsorption (Brunauer, Emmett, and Teller (B.E.T.)] method at very low temperatures. The pore diameter distribution can also be found from adsorption–desorption versus pressure curves, or by high-pressure mercury

Table 8.6 Surface Area, Pore Volume, and Mean Pore Diameter for Adsorbent Materials[a]

Material	Area ($m^2\ g^{-1}$)	Pore volume ($cm^3\ g^{-1}$)	Mean pore diameter (10^{-10} m)
Activated carbon	500–1,500	0.6–0.8	20–40
Silica gels	200–600	~0.4	30–200
Activated alumina	175	0.39	90
Kieselguhr	4.2	1.14	22,000

[a] From Wheeler (1951).

injection prosimetry. Some typical data for adsorbent materials is shown in Table 8.6.

The adsorption of different vapors on the same adsorbent is characterized by different adsorption isotherms giving the grams of adsorbed vapor (adsorbate) per gram of adsorbent as a function of the equilibrium partial pressure of the vapor. An example of adsorption isotherms (physical adsorption) is shown in Fig. 8.10. It is evident from the figure that the amount of adsorption decreases with increasing temperature.

Figure 8.10 Effect of temperature on the adsorption of ethyl chloride on charcoal (Goldmann and Polanyi, 1928).

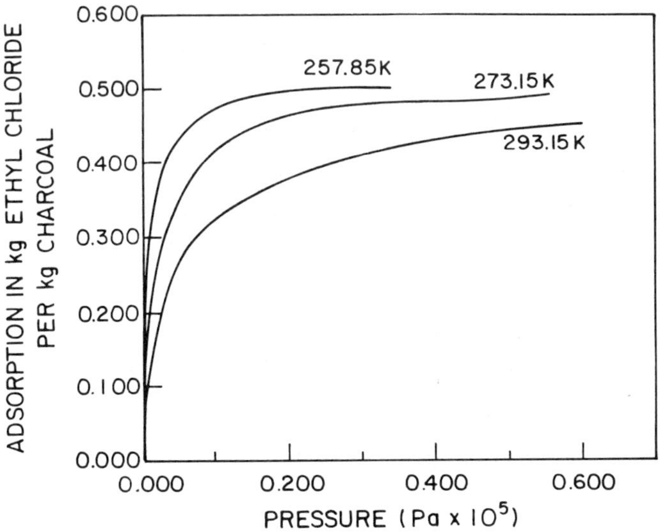

Activated carbon is used for the recovery of hydrocarbon solvents, the removal and recovery of hydrocarbon gases from coal gas and natural gas, and the removal of odors and other trace impurities from gas streams. In the first two cases, the recovery of hydrocarbons is economically justified, as the products are of value and can be used commercially. The plant consists of two deep beds of carbon housed in pressure vessels that are alternatively charged with the hydrocarbons and then steamed at 100–105°C to remove the adsorbed material. The control of odors with active charcoal is very effective. Thin beds with flow rates and a life of 12–18 months or even longer are used continuously before replacement becomes necessary. Among the advantages of adsorption on nonpolar adsorbents is the possibility of recovering valuable pollutants after adsorption, and the very low concentrations to which adsorption is applicable.

Polar Adsorbents

Polar adsorbents are either silica or metallic oxides. The siliceous materials are silica gel, Fuller's earth, diatomaceous earths, such as kieselguhr, and synthetic and natural zeolites. These materials have an affinity for polar as well as nonpolar molecules but will adsorb polar molecules in preference to nonpolar ones. Metallic oxides, such as aluminium oxide in the form of activated alumina or activated bauxite, have an even higher attraction for polar molecules. Therefore, these materials are commonly used for the removal of water vapor, rather than organic molecules, from gas streams. The synthetic zeolites, sometimes called molecular sieves, are sodium or calcium aluminosilicates, activated by heating, which drives off the water of crystallization. The chief advantage of the molecular sieves is that they can be used for drying at elevated temperatures, where silica gel and alumina lose their effectiveness. Another application is the use of these adsorbents for the selective adsorption of polar molecules, such as water, carbon dioxide, ammonia, acetylene, hydrogen sulfide, and sulfur dioxide. They are, therefore, used for the purification of inert gases, the removal of carbon dioxide and water from ethylene before subjecting this to polymerization for polyethylene, and the purification of natural gas.

Adsorption with Chemical Reaction

Adsorbent surfaces that react chemically with the gas molecules, changing these to more useful or desirable ones before release, open the whole field of surface reactions and catalysis. Only one example of particular interest in air pollution control and gas purification is given here. As in the case of adsorption without chemical reaction, it is necessary to have a large surface area of reagent accessible to the gas.

Important among these processes are the dry oxidation processes for the removal of sulfur compounds from coal gas (Strauss, 1975). Chemically, the hydrogen sulfide is converted to sulfur, using oxygen in carriers that react readily with it at ordinary temperatures. The most important of these is ferric oxide, and the basic chemistry of the process can be written as a reaction and regeneration stage:

$$6H_2S + 2Fe_2O_3 \rightarrow 2Fe_2S_3 + 6H_2O \qquad \text{Reaction}$$

$$2Fe_2S_3 + 3O_2 \rightarrow 6S + 2Fe_2O_3 \qquad \text{Regeneration}$$

Only the α and γ forms of ferric oxide are reactive, and the reaction proceeds best at a temperature of 37.8°C and in an alkaline environment. The iron oxide is used in a finely divided form either mixed with peat or fibrous matter, containing 40% water and with a porosity of about 60%. Alternatively, the iron oxide can be supported on a large surface area, such as wood shavings or crushed slags, and the alkalinity is carefully controlled to pH 8–8.5.

Adsorption Equipment

Fixed Regenerable Beds

Usually, it is not economical to throw away the adsorbent after it has been once saturated with the adsorbate, and therefore, it is customary to regenerate the adsorbent beds. Regeneration is done mostly by heating, often by steaming. In thin beds, usually a few inches thick, the adsorbent is held between vertical screens or perforated metal sheets, and the gas passes through horizontally. Thin beds are characterized by very low pressure drops and low adsorption capacity. A thick (deep) bed train is shown diagrammatically in Fig. 8.11. The adsorbent is contained in long horizontal tanks. A minimum of two identical deep beds is necessary, one adsorbing the pollutant while the other one is being regenerated. The four units shown in the figure are used cyclically in the sequences 1234, 2341, 3412, 4123, and back to 1234.

The principle of operation of a single adsorbent bed is shown in Fig. 8.12. The curves in the figure, called adsorption waves, show the adsorbate concentration profiles within the bed at various points of time. Curve A is the concentration profile shortly after starting to pass the gas containing pollutant through the freshly regenerated bed. It is evident that the inlet face of the bed is already saturated with the absorbate, while most of the bed does not contain any adsorbate yet. Curves B and C show the concentration profiles at increasing points of time. Curve D is the breakthrough curve, representing the point of time when the pollutant first

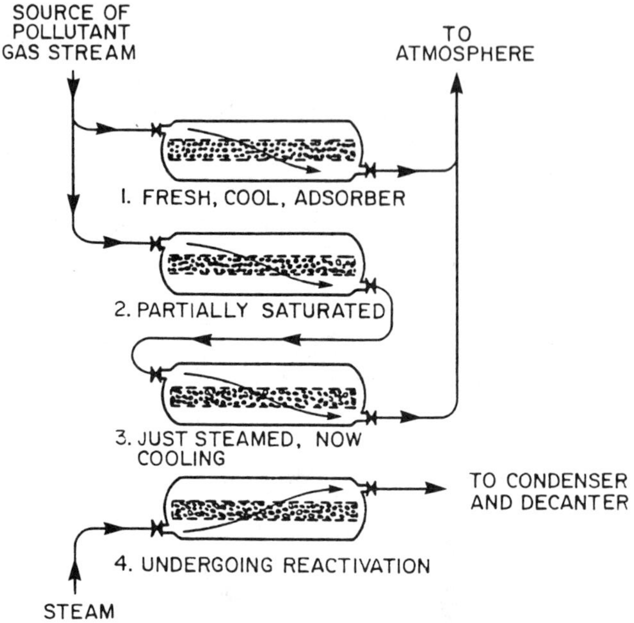

Figure 8.11 Four-stage regenerative adsorber (Turk, 1977).

Figure 8.12 Propagation of an adsorption wave (Turk, 1977).

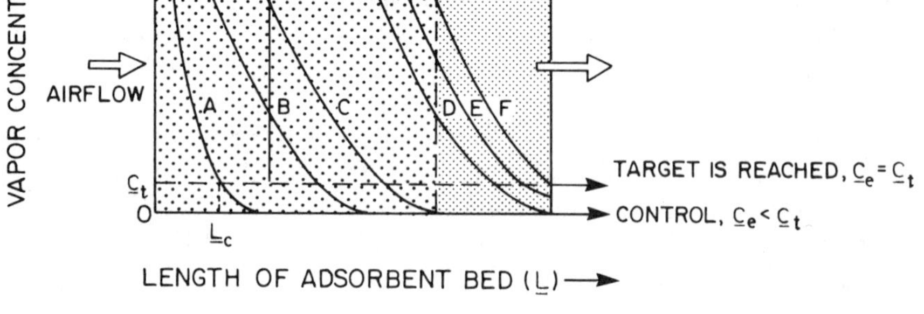

appears at the outlet face of the bed. After that moment, the pollutant concentration in the gas emerging from the bed gradually increases, as indicated by curves E and F. Curve F corresponds to the maximum pollutant concentration that is tolerated, and at that point the bed is taken off the stream for regeneration. In the figure, L_c is the minimum bed depth that could be used to reach the target concentration C_t. The section of the bed between complete saturation and zero saturation is the transfer zone in which adsorption is taking place. The length of the transfer zone increases with time from zero to T_c, corresponding to curve C, and after that it remains constant until breakthrough (curve D), that is, $T_d = T_c$. At this point, the upstream section of the bed S_d is completely saturated and, hence, inactive.

In systems like the one shown in Fig. 8.11, the entire sorbent bed is permitted to be saturated before it is regenerated. The duration of service t (h), before regeneration, at 20°C and 760 Torr, of carbon sorbent beds has been expressed by the formula (Turk *et al.*, 1952)

$$t = 6.43 \times 10^6 SW/EQMC \tag{8.27}$$

where S is the retentivity (weight of adsorbate in completely saturated bed × 100/weight of adsorbent; see Table 8.7), W is the weight of adsorbent (pounds), E is the sorption efficiency (fraction), Q is the airflow rate through the sorbent bed (ft^3/min), M is the average molecular weight

Table 8.7 Retentivity of Vapors by Activated Carbon[a, b]

Substance	Formula	Molecular weight	Normal boiling point (°C)	Approximate retentivity[a] (% at 20°C, 760 Torr)
Acetaldehyde	C_2H_4O	44.1	21	7
Amyl acetate	$C_7H_{14}O_2$	130.2	148	34
Butyric acid	$C_4H_8O_2$	88.1	164	35
Carbon tetrachloride	CCl_4	153.8	76	45
Ethyl acetate	$C_4H_8O_2$	88.1	77	19
Ethyl mercaptan	C_2H_6S	62.1	35	23
Eucalyptole	$C_{10}H_{18}O$	154.2	176	20
Ozone	O_3	48.0	−112	Decomposes to oxygen
Putrescine	$C_4H_{12}N_2$	88.2	158	25
Skatole	C_9H_9N	131.2	266	25
Toluene	C_7H_8	92.1	111	29

[a] Perecentage retained in a dry airstream at 20°C, 760 Torr, based on weight of carbon.
[b] From Turk *et al.* (1952).

of adsorbate vapor (g/mol), and C is the entering adsorbate vapor concentration (ppm by volume).

Traveling Bed Adsorbers

These are continuous countercurrent column adsorbers. Freshly regenerated adsorbent is added at the top of the column, and the adsorbent travels downward to the bottom of the column. Gas to be treated enters at the bottom, passes up the column, and leaves at the top. In addition, there are a few examples of fluidized bed adsorber systems in use.

8.5 Combustion

Combustion Stoichiometric Balance

Combustion is a chemical reaction of combustibles with oxygen, forming carbon dioxide, CO_2, and water, H_2O. In incomplete combustion, carbon monoxide, CO, is also formed. Combustion is, in principle, always incomplete in another sense, that is, not 100% of the combustibles are reacted. First, the temperature must be sufficiently high for the combustion to take place at all or to occur fast enough. Second, sufficient contacting between the combustibles and the oxygen must be ensured for a sufficiently long time to approach complete combustion. Good contacting requires highly turbulent flow conditions of the air. Below a certain concentration of either the oxygen or the combustible, or both, the combustion reaction cannot proceed. All of these limitations on the extent of combustion concern the rate of combustion; therefore, they are of a kinetic nature. They must be taken into account in the design of the burner or combustor. In principle, the maximum possible extent of combustion is the equilibrium conversion of the combustibles that may be reached when the time factor is irrelevant, that is, in infinitely long time. In combustion reactions, the equilibrium conversion is usually very near 100%, even though, strictly speaking, 100% conversion is theoretically impossible. In general, with greater amounts of oxygen used, more of the combustibles will be combusted. This statement, however, does not mean simply that unless enough oxygen is supplied the combustibles cannot be completely combusted. It means, far less obviously, that an excess of oxygen helps the combustion both to take place faster and to go closer to completion.

In order to understand this, let us first write down the general stoichiometric equation of combustion of a hydrocarbon, C_mH_n:

$$C_mH_n + \left(m + \frac{n}{4}\right)O_2 \rightarrow mCO_2 + \left(\frac{n}{2}\right)H_2O$$

This equation is merely a material balance, where the chemical formulas mean mass quantities of C_mH_n, O_2, CO_2, and H_2O, which are equal or proportional to the corresponding molecular weight of these substances (e.g., 32 for O_2). For example, for $m = 1$, $n = 4$ (CH_4, methane), it means that *if* 16 g of CH_4 react with 64 g of O_2, then 44 g of CO_2 and 36 g of H_2O are formed.

When the combustible gas and oxygen are contacted in the proportions indicated by the stoichiometric equation (material balance), then they are said to be present in stoichiometric or theoretical proportions. It is generally true that if more oxygen is supplied than required by the stoichiometric equation, for example, 128 g of O_2 for 16 g of CH_4, that is, a 100% excess over the stoichiometric proportions, then everything else being unchanged, a larger perecentage of the CH_4 (and a correspondingly lower percentage of O_2) will react when equilibrium is reached. The amounts reacted, however, will remain in the stoichiometric proportions indicated by the stoichiometric formula.

The equilibrium conversion is calculated from the equilibrium constant of the reaction K_p, defined as

$$K_p = p_{CO_2}^m p_{H_2O}^{(n/2)} / p_{C_mH_n} p_{O_2}^{(m+n/4)} \tag{8.28}$$

where all the components are in the gaseous state, and p_i is the partial pressure of the gas i.

The partial pressures are related to the fractional equilibrium conversion ξ as now explained. Let us assume that at the start of the combustion process there were these numbers of moles present of each gas in the reaction mixture: $n_{C_mH_n}^0$, $n_{O_2}^0$, $n_{CO_2}^0$, and $n_{H_2O}^0$ (the last two are usually present in negligibly small quantities). At equilibrium, the mole numbers present in the gas mixture are

$$n_{C_mH_n} = n_{C_mH_n}^0 - \xi n_{C_mH_n}^0 = n_{C_mH_n}^0 (1 - \xi)$$

$$n_{O_2} = n_{O_2}^0 - \left(m + \frac{n}{4}\right)\xi n_{O_2}^0 = n_{O_2}^0\left[1 - \left(m + \frac{n}{4}\right)\xi\right]$$

$$n_{CO_2} = n_{CO_2}^0 + m\xi n_{CO_2}^0 = n_{CO_2}^0 (1 + m\xi)$$

$$n_{H_2O} = n_{H_2O}^0 + \frac{n}{2}\xi n_{H_2O}^0 = n_{H_2O}^0\left(1 + \frac{n}{2}\xi\right)$$

There are usually also n_{N_2} moles of nitrogen present that react negligibly. Whence, the total number of moles present in equilibrium is ($n_{N_2} + n_{C_mH_n}$

$+ n_{O_2} + n_{CO_2} + n_{H_2O}) = \Sigma n_i$. The individual mole fractions at equilibrium are

$$x_{C_mH_n} = \frac{n^0_{C_mH_n}(1 - \xi)}{\Sigma n_i}; \qquad x_{O_2} = \frac{n^0_{O_2}\{1 - [m + (n/4)]\xi\}}{\Sigma n_i}$$

$$x_{CO_2} = \frac{n^0_{CO_2}(1 + m\xi)}{\Sigma n_i}; \qquad x_{H_2O} = \frac{n^0_{H_2O}[1 + (n/2)\xi]}{\Sigma n_i}$$

As each partial pressure p_i is related to the corresponding mole fraction x_i and the total pressure p by the relation $p_i = px_i$, the equilibrium constant may be written as

$$K_p = p^{\{m+(n/2)-1-[m+(n/4)]\}} \frac{x^m_{CO_2} x^{(n/2)}_{H_2O}}{x_{C_mH_n} x^{(m+n/4)}_{O_2}} \tag{8.29}$$

In the knowledge of K_p, p, m, and n, this equation is solved for ξ, the equilibrium conversion. The equilibrium constant K_p is usually calculated from the standard Gibbs energy of the reaction, ΔG^0, obtained from tables or graphs, as

$$K_p = \exp(-\Delta G^0/RT) \tag{8.30}$$

where R is the universal gas constant and T is the absolute temperature in kelvin (see, for example, Castellan, 1983). As the value of $-\Delta G^0$ for combustion reactions is on the order of several hundred kJ/mol, the value of K_p is usually in the range between 10^{10} and 10^{40}, which is tantamount to complete equilibrium conversion.

Theoretical Combustion

As explained in the previous section, no combustion is truly 100%. Nevertheless, it is convenient, for simplicity, to introduce the concept of *theoretical combustion*, meaning that the combustion reaction is assumed to be 100%, as indicated by the stoichiometric equation. For example, the stoichiometric combustion equation of methane with air is

$$CH_4 + 2O_2 + 7.52N_2 \rightleftharpoons CO_2 + 2H_2O + 7.52N_2$$

Theoretical combustion means that if 16 g of methane are burned in 64 g of oxygen and 210 g of N_2, then all the 16 g of methane is assumed to be combusted (and all of the 64 g of the oxygen used up). If, for example, only 32 g of oxygen were used, then in theoretical combustion, 8 g of the 16 g of methane present would be combusted. Further, in 67.2 g of oxygen and 221 g of nitrogen, representing 5% excess air, all the 16 g of methane present is

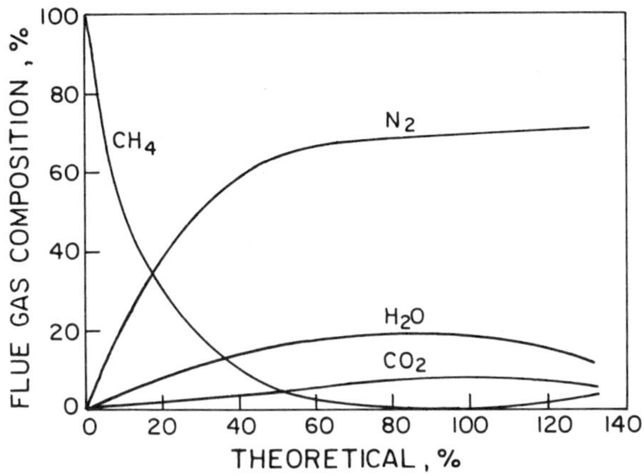

Figure 8.13 Flue gas composition for theoretical combustion of natural gas with air (Hesketh, 1979).

assumed to be combusted and, as only 64 g of oxygen can react with 16 g of methane, 3.2 g of oxygen remains unreacted. The flue gas composition for different amounts of air used in the (theoretical) combustion of methane (natural gas) can be readily calculated; the results of such calculations are shown in Fig. 8.13.

Definitions

1. The air–fuel ratio (AF) is the mass of air per unit mass of fuel.
2. The flue gas quantity is the volume of flue gas per mol of fuel. This can be reported either in actual cubic meters (or cubic feet) on a wet basis or in standard cubic meters (or cubic feet) on a dry basis.

EXAMPLE 8.3

Calculate the flue gas composition for the theoretical combustion of natural gas (methane) in

1. a theoretical amount of air,
2. 5% excess air,
3. a dry gas composition in case 2,
4. the air–fuel ratio in case 2, and
5. the flue gas quantity in case 2.

SOLUTION

1. Air is considered to be 79% N_2 (by volume or by mole numbers) and 21% O_2. The total number of moles in the flue gas is

$$1 + 2 + (\tfrac{79}{21} \times 2 = 7.52) = 10.52 \text{ mol}$$

Hence,

$$(1/10.52)(100) = 9.5 \text{ mol or vol \% } CO_2$$

$$(2/10.52)(100) = 19.0 \text{ mol or vol \% } H_2O$$

$$(7.52/10.52)(100) = 71.5 \text{ mol or vol \% } N_2$$

Since the water is included in the gas composition, this is called wet gas analysis.

2. The total number of moles in the flue gas is

$$(1 + 2 + 0.1 + 7.9) = 11.00 \text{ mol}$$

Hence, the theoretical composition of the flue gas is

$$(1/11)(100) = 9.09 \text{ mol or vol \% } CO_2$$

$$(2/11)(100) = 18.18 \text{ mol or vol \% } H_2O$$

$$(0.1/11)(100) = 0.91 \text{ mol or vol \% } O_2$$

$$(7.9/11)(100) = 71.82 \text{ mol or vol \% } N_2$$

3. Dry gas composition:

$$\left(\frac{9.09}{100 - 18.18}\right)(100) = 11.11 \text{ mol or vol \% } CO_2$$

$$\left(\frac{0.91}{100 - 18.18}\right)(100) = 1.11 \text{ mol or vol \% } O_2$$

$$\left(\frac{71.82}{100 - 18.18}\right)(100) = 87.78 \text{ mol or vol \% } N_2$$

4. The average molecular weight of air is

$$(21 \times 36 + 79 \times 28)/100 \approx 29$$

Hence,

$$\text{AF} = \frac{2.1 \text{ mol } O_2 \text{ (100 mol air/21 mol air } O_2\text{) (29 g air/mol air)}}{1 \text{ mol } CH_4 \text{ (16 g } CH_4\text{/mol)}}$$

$$= 18.1 \left(\frac{\text{g air}}{\text{g fuel}}\right)$$

5. Assuming a flue gas temperature of 160°C, one has on a wet basis,

$$\left(\frac{11 \text{ mol flue gas}}{1 \text{ mol } CH_4}\right)\left(\frac{22.4 \text{ liters}}{\text{mol gas}}\right)\left(\frac{273 + 160}{273}\right)\left(\frac{1 \text{ m}^3}{1000 \text{ liters}}\right) = 0.391 \frac{\text{Nm}^3}{\text{mol } CH_4}$$

and on a dry basis,

$$\left(\frac{9 \text{ mol flue gas}}{1 \text{ mol CH}_4}\right)\left(\frac{22.4 \text{ liters}}{\text{mol gas}}\right)\left(\frac{273 + 20}{273}\right)\left(\frac{1 \text{ m}^3}{1000 \text{ liters}}\right) = 0.216 \frac{\text{Nm}^3}{\text{mol CH}_4}$$

8.6 Different Kinds of Fuels

Gases

In most cases, natural gas consists almost entirely of methane, CH_4. It also contains small amounts of ethane, helium, and carbon dioxide, and traces of H_2S and mercaptans. Liquified petroleum gas (LPG) consists mostly of propane and butane. It is popularly known as bottled gas. Artificial, or city gas, carburated (water) gas, and producer and coke oven gas are manufactured from coal or coke. They all contain a high perecentage of hydrogen and the highly poisonous carbon monoxide, CO. Gases are the cleanest burning fuels and enjoy great popularity. About 25% of the energy in North America is derived from gases. Typical uncontrolled emissions from gas combustion are shown in Table 8.8.

Table 8.8 Typical Uncontrolled Emissions from Gas Combustion[a]

		Uncontrolled emissions[b]				
Boiler	Fuel	Particulate matter	Sulfur oxides[c]	Carbon monoxide	Hydro-carbons[d]	Nitrogen oxides[e]
Utility or large industrial	Natural gas	80–240	20.9	272	16	11,200[f]
Commercial or industrial	Natural gas	80–240	20.9	272	48	1,920–3,680
	Butane (LPG)	0.22	0.01	0.19	0.036	1.45
	Propane (LPG)	0.20	0.01	0.18	0.036	1.35
Small commercial or domestic	Natural gas	80–240	20.9	320	128	1,280–1,920[g]
	Butane (LPG)	0.23	0.01	0.24	0.096	1.0–1.5[g]
	Propane (LPG)	0.22	0.01	0.23	0.084	0.8–1.3[g]

[a] From EPA (1973–1978).
[b] Reported as $g/10^3$ m^3 natural gas or g/liter LPG.
[c] Multiply SO_2 value listed times g sulfur/100 Nm^3 natural gas or times g sulfur/100 m^3 LPG vapor; pollutant weight reported as SO_2.
[d] Pollutant weight reported as CH_4.
[e] Pollutant weight reported as NO_2.
[f] Multiply NO_2 value by $0.151 \exp(-0.0189L)$, where L is percentage load on boiler; use 4,800 $g/10^3$ m^3 for tangentially fired units.
[g] Lower value is for domestic and higher is for commercial heating systems.

Table 8.9 Typical Uncontrolled Emissions from Oil Combustion[a]

		Uncontrolled Emissions (g/liter of fuel)					
Boiler	Fuel oil	Particulate matter	Sulfur oxides[b]	Carbon monoxide	Hydro-carbons[c]	Nitrogen oxides[d]	Aldehydes
Utility or large industrial	Residual	1	19.2 S	0.4	0.25	12.6	0.12
Commercial or industrial	Residual	2.75	19.2 S	0.5	0.35	9.6	0.12
	Distillate	1.8	17.2 S	0.5	0.35	9.6	0.25
Small commercial or domestic	Distillate	1.2	17.2 S	0.6	0.35	1.5	0.25

[a] From EPA (1973–1978).
[b] S is the weight percentage of sulfur in oil; pollutant weight is reported as SO_2.
[c] Pollutant weight is reported as CH_4.
[d] Pollutant weight is reported as NO_2; use one-half this value for tangentially fired boilers.

Oil

Fuel oils are mostly obtained in refineries from crude oil. However, processes do exist that produce synthetic oil from coal or refuse. Oil is burned in the form of small (about 50 μm) drops, mixed with air, which are produced by atomizing nozzles. Emissions from the combustion of oil are less clean than those from gas. Typical uncontrolled emissions from horizontally fired oil furnaces are shown in Table 8.9.

EXAMPLE 8.4

A fuel oil of specific gravity 0.90 contains 87% by weight carbon, 11.5% by weight hydrogen, and 2% by weight sulfur. The dry flue gas composition is 11.36% CO_2, 5.98% O_2, 0.10% SO_2, and 82.56% N_2. Calculate the percentage of excess air and the flue gas composition on a wet basis.

SOLUTION

The theoretical air needed is

$$C + O_2 \rightarrow CO_2, \qquad H_2 + \tfrac{1}{2}O_2 \rightarrow H_2O, \qquad S + O_2 \rightarrow SO_2$$

Basis is 100 g of fuel oil. This contains

$$87/12 = 7.25 \text{ mol C}, \qquad 11.5/2 = 5.75 \text{ mol } H_2$$

and

$$2/32 = 0.0625 \text{ mol S}$$

As all of the fuel is assumed to have burned, this corresponds to 7.25 mol CO_2, 5.75 mol H_2O, and 0.0625 mol S. Hence, the theoretical amount of O_2 to burn 100 g fuel oil is 7.25 + (5.75/2) + 0.625 = 10.1875 mol O_2.

The excess amount of O_2 used is obtained in two steps. First, the excess amount present in 100 mol dry flue gas is 5.98 mol O_2. Second, based on 100 g fuel oil, this gives

$$5.98 \frac{\text{mol } O_2}{100 \text{ mol dry flue gas}} \times \frac{7.25 \text{ mol C}/100 \text{ g fuel oil}}{11.36 \text{ mol C}/100 \text{ mol dry flue gas}}$$

$$= 3.82 \frac{\text{mol } O_2}{100 \text{ g fuel oil}} = \text{excess oxygen}$$

The percentage of excess oxygen (air) is then

$$(3.820/10.1875) \times 100 = 37.5\%$$

The wet gas analysis is obtained by first calculating the number of moles of H_2O produced per 100 mol of dry flue gas:

$$5.75 \frac{\text{mol } H_2O}{100 \text{ g fuel oil}} \times \frac{11.36 \text{ mol C}/100 \text{ mol dry flue gas}}{7.25 \text{ mol C}/100 \text{ g fuel oil}}$$

$$= 9.01 \frac{\text{mol } H_2O}{100 \text{ mol dry flue gas}}$$

Then,

$$\frac{11.36 \times 100}{109.01} = 10.42\% \ CO_2; \qquad \frac{5.98 \times 100}{109.01} = 5.49\% \ O_2; \qquad 0.10\% \ SO_2$$

$$\frac{82.56 \times 100}{109.01} = 75.74\% \ N_2; \qquad \frac{9.01 \times 100}{109.01} = 8.27\% \ H_2O$$

Coal

One distinguishes anthracite, bituminous, and subbituminous coal and lignite in decreasing order of heat value. When coal is heated, volatile substances, on the order of 20% by weight escape, leaving behind about 75% by weight of fixed carbon and inorganic matter or ash. Coal also contains sulfur in three different forms, that is, as pyrites, organic sulfur, and sulfates.

Coal is burned in piled beds, by suspension firing, or in fluidized beds. Coal-bed firing is usually done in underfeed, crossfeed, or overfeed strokers. Suspension firing uses crushed or pulverized coal. Fluidized-bed combustion is made very suitable for SO_2 removal by adding limestone particles to the bed.

The combustion products of coal contain, in addition to CO_2 and H_2O, CO, SO_2, NO_x, hydrocarbons, aldehydes, and fly ash. Also present are trace amounts of arsenic, beryllium, lead, mercury, and radioactive materials. These are hazardous and toxic, presenting both health problems and difficulties in choosing the material of construction for air pollution control devices. Typical uncontrolled emissions from bituminous coal combustion are given in Table 8.10.

Table 8.10 Typical Uncontrolled Emissions from Bituminous Coal Combustion[a]

Furnace			Uncontrolled emissions (g/kg coal)					
Boiler category	Heat input (10^6 kcal/h)	Type	Particulate matter[b]	Sulfur oxides[c]	Carbon monoxide	Hydro-carbons[d]	Nitrogen oxides[e]	Aldehydes
Utility or large industrial	> 25	Pulverized	8 A	19 S	0.5	0.15	9	0.0025
		cyclone	1 A	19 S	0.5	0.15	27.5	0.0025
Commercial or industrial	2.5–25	Underfeed and crossfeed	2.5 A	19 S	1	0.5	7.5	0.0025
		Overfeed	6.5 A	19 S	1	0.5	7.5	0.0025
Small commercial or domestic	< 2.5	Overfeed	1 A	19 S	5	5	3.0	0.0025
		Hand fired	20 A	19 S	45	45	1.5	0.0025

[a] From EPA (1973–1978).
[b] A is the weight percentage ash in coal.
[b] S is the weight percentage sulfur in coal; pollutant weight is reported as SO_2.
[d] Pollutant weight is reported as CH_4.
[e] Pollutant weight is reported as NO_2.

EXAMPLE 8.5

Estimate the flue gas composition on a wet basis and the quantity in A m^3/h at 160°C from a 100-MW boiler. The coal analysis is 80.00% C, 5.00% H_2, 0.8% S, 1.5% N_2, 6.5% ash, and 6.2% O_2 by difference. Assume that the boiler operates at 27% excess air and the heat value of the coal is 7.0 kcal/g coal. Assume theoretical combustion.

SOLUTION

On the basis of 100 g of coal:

$$(80/12) = 6.67 \text{ mol C}$$

$$(5/2) = 2.50 \text{ mol } H_2$$

$$(0.8/32) = 0.025 \text{ mol S}$$

$$(1.5/28) = 0.054 \text{ mol } N_2$$

$$(6.5/32) = 0.203 \text{ mol } O_2$$

Theoretical combustion requires

$$6.67 + (2.50/2) + 0.025 - 0.203 = 7.74 \text{ mol } O_2$$

that is,

$$7.74(100/21) = 36.87 \text{ mol air}$$

Using 27% excess air,

$$36.87(1.27) = 46.82 \text{ mol air}$$

Total number of moles in flue gas is

$$6.67 + 2.50 + 0.025 + 0.054 + (46.82)(0.79) + (7.74)(0.27) = 48.33 \text{ mol}$$

giving the following flue gas composition

$$(6.67/48.33)100 = 13.80 \text{ vol. \% } CO_2$$

$$(2.50/48.33)100 = 5.17 \text{ vol. \% } H_2O$$

$$(0.025/48.33)100 = 0.05 \text{ vol. \% } SO_2$$

$$(37.04/48.33)100 = 76.64 \text{ vol. \% } N_2$$

$$(2.09/48.33)100 = 4.32 \text{ vol. \% } O_2$$

The volume of flue gas is calculated as

$$\left(\frac{48.33 \text{ mol gas}}{100 \text{ g coal}}\right)\left(\frac{19 \text{ coal}}{7 \text{ kcal}}\right)\left(\frac{2.5 \times 10^8 \text{ kcal/h}}{100 \text{ MW}}\right)(100 \text{ MW})$$

$$\times\left(\frac{22.4 \text{ liters}}{\text{mol gas}}\right)\left(\frac{273 + 160}{273}\right)\left(\frac{1 \text{ m}^3}{10^3 \text{ liters}}\right) = 585{,}633 \text{ A m}^3/\text{h}$$

which is equivalent to 369,233 Nm^3/h. The reference literature (Smith and Gruber, 1966) suggests that the flue gas quantity under the conditions of this problem should be equal to about 340,000 Nm^3/h, which is close enough.

8.7 Combustion Processes for the Removal of Air Pollutants

In many industrial processes, gases or particulates are emitted into the atmosphere that are combustible. Solvent evaporation processes are responsible for about one half of this kind of pollutant. Additional contributors include petroleum refining, gasoline marketing, fat rendering, and coffee roasting (Rolke *et al.*, 1972). Combustible pollutants are currently destroyed commercially by means of one of these four devices: thermal afterburners, catalytic afterburners, furnaces, or flares (Gottschlich, 1977).

The products of combustion are carbon dioxide and water in the case of organic material or sulfur dioxide and water from organic sulfides. If the concentration of the gas is sufficiently high to be in the flammable range, the oxidation process can be self-supporting after the mixture is ignited. The lowest vapor concentration at which this occurs is the lower explosion (flammable) limit (LEL), while the maximum concentration is the upper explosion (flammable) limit (UEL). Within this range, controlled combustion can be achieved, but under certain circumstances explosions also occur.

The temperature above which combustion of the vapor and gases is sustained is the autogenous combustion temperature. The flammable range depends primarily on the amount of energy that is released by the combustion process, that is, the available net heat of combustion per standard cubic meter of gas (J m^{-3} at 21°C). The lower flammable limit is approximately 1.9 MJ m^{-3} (21°C), and fume energy concentrations higher than this are necessary for self-supporting flames; while for a flame with good combustion characteristics, a fume energy concentration of about 3.7 MJ m^{-3} (21°C) is desirable.

Thermal Afterburners

Thermal afterburners are used when the combustible concentration is below the LEL. It usually consists of a combustion chamber equipped with a burner at the gas inlet end, and it is operated in the temperature range 550°C to 850°C. Supplementary air may or may not be added, depending on the need for it. About 15–20% excess oxygen is needed over the stoichiometric quantity. Mostly gas is used as fuel for the burner. Once the gases have been heated up by the flame above 700°C, the oxidation continues in the rest of the chamber where there is no burner or flame. The temperature in the combustion chamber or in the flame must be sufficiently high, there must be adequate turbulent mixing of the oxygen and the combustible gases, and sufficient time for combustion has to be available. The design of the burner and combustion chamber and the degree of premixing of gases determine these factors. A premixed flame tends to be shorter, hotter, and blue in color, while a nonpremixed flame tends to be luminous because of the cracking of hydrocarbon vapors, which produces luminous carbon particles.

Catalytic Afterburners

Catalytic combustion is an extension of the technique of using a combustion chamber with the advantage that oxidation on the catalyst surface will occur well below the autogenous combustion temperature, and also in gas concentrations below the fume concentration energy required for self-supporting combustion processes. An additional advantage is that only the stoichiometric quantity of oxygen (as air) has to be provided, minimizing the preheating requirements. Catalytic afterburners usually operate in the 400–500°C temperature range.

Catalytic afterburners usually consist of a preheater section, containing either a burner or a heat exchanger, followed by a bed of catalyst, such as platinum or certain metal oxides, supported on a solid carrier. Catalysts are expensive, but the amount of catalyst used is small so that the purchase cost of a catalytic afterburner is competitive with the cost of the thermal afterburner. Catalytic combustion requires only the stoichiometric quantity of oxygen, and so the fuel requirement is also reduced. Fuel can also be saved by using a heat exchanger to transfer heat from the exit stream to the inlet stream.

Normally, any gas-borne organic material can be treated by catalytic combustion provided that the combustion products are gaseous. Therefore, organic sulfur and nitrogen compounds can generally be treated, but organic silicones and phosphates must be excluded because of the nature of the combustion products. If dusts or inorganic materials are present in an appreciable quantity, they should be removed; but small quantities such as are normally found in the air will pass through the catalytic combustion unit or occasionally be caught in it. These residues can be removed by occasional (annual or semiannual) washing of the elements.

Furnaces

Sometimes the already existing process furnace can be used to dispose of the combustible pollutants. Such methods are limited by the fact that the additional gas stream may interfere with the operation of the main process.

Flares

The purpose of flaring is to get rid of the combustible gases that are not needed. A flare is usually mounted high above the ground level, but sometimes ground-level flares are also used. Essentially, a flare is a safe, high-efficiency burner that ensures complete combustion despite rapid changes in the flow rate of the gas stream. Flares are normally nozzle mix burners. Steam jets are frequently used to ensure good mixing and an adequate flow of combustion air.

For the flaring of low-value waste gases, the usual practice is to use high-velocity steam jets surrounding the gas jet, although this leads to rather loud flares on occasions. Alternative types of flares using premixing with air, on the bunsen burner principle, or premixing with steam with subsequent entrainment of air at the jet are also used.

8.8 Odor Control

Odor is the most irritating form of air pollution. About half of all the complaints by residents to local pollution control agencies are associated with odors (Leonards, 1984). Odor is a sensation resulting from the interaction of certain chemical species (odorants) with the olfactory area located in the upper reaches of the nose. Studies of the odor properties of chemicals have indicated odor threshold concentrations of 0.000001 ppm for *o*-iodophenol and *o*-bromophenol; the recognition odor threshold concentration for methylene chloride is 214 ppm (Leonards, 1984). Odors may be released from stacks or vents, leaks, spills, flares, wastewater treatment, sludge handling processes, feed lots, etc.

Odor problems are defined essentially in three steps: the ambient survey, on-site review, and source sampling and evaluation. The purposes of the ambient odor survey are to determine the quality and intensity of odors in the areas surrounding the facility and to study their pattern of occurrence. The on-site review is best carried out in conjunction with the ambient odor survey in order to be able to relate the odors observed off-site with particular sources within the plant. The on-site review should cover all the operations and all the locations in the plant. After the sources of odor have been identified, samples are taken from these sources in an attempt to quantify the odor emission levels and to determine the degree of control necessary.

Much of the sampling in the past has been based on the dilution to threshold concept, according to which the odor-bearing air is sampled in glass bulbs or bags and then diluted in an olfactometer with various amounts of odor-free air to the "threshold" of an observer. Usually, a panel of observers has been used, because of the varying sensitivity of individuals to odors. Recently, the Chromosorb 102 source sampling technique was developed by Arthur D. Little, Inc. Chromosorb 102 is placed in a stainless steel sorbent trap that is part of a sampling train. After sampling, the odorant species are quantitatively eluted with high-purity pentane. For sensory evaluation, the sample is tested with the help of an olfactometer in which known dilutions of the sample are prepared. The advantage of the Chromosorb sampling procedure is that large volumes of the emission can be collected. More details of odor analysis can be found in Leonards (1984).

Once the source of the odor has been identified, elimination of the odor may be achieved by removing the particulates if the odor is associated with the emission of some particulate pollutant; adsorping the odor by active carbon; or scrubbing with aqueous potassium permanganete solution if the odor is associated with the emission of an organic gas or vapor. Large area odors, associated with activities such as the spreading of manure in the

Table 8.11 Common Industrial Odor Sources[a]

Odor source	Cause of odors
1. Pulp and paper manufacturing	Hydrogen sulfide, organic sulfur compounds
2. Plastics industries	Phenols, formaldehyde, acrolein, organic sulfur compound, organic solvents
3. Fat- and oil-based industries	Aldehydes, amines,
4. Waste treatment facilities	Organic nitrogen compounds, hydrogen sulfide, organic sulfur compounds
5. Rendering and related industries	Aldehydes, ketones, organic acids, organic sulfur compounds, amines
6. Food canning and processing	Aldehydes, amines, other nitrogenous compounds
7. Pharmaceutical industry	Organic sulfur compounds, amines
8. Textiles	Starch, formaldehyde

[a]From Cheremisinoff and Young (1975).

fields, for example, have no known methods of control. Common industrial odor sources are listed in Table 8.11.

PROBLEMS

1. Repeat the problem worked in Example 8.1, but use 1.2 times the minimum oil circulation rate and twice the minimum oil circulation rate. Discuss your results.
2. Calculate the tower diameter for the data in problem 1 for a 1 cm H_2O pressure drop per meter of packing height. Assume 2.5-cm ceramic Raschig rings.
3. Calculate the tower height for the data in problem 1 and problem 2, assuming $K_G a = 0.4$ mol benzene/h cm^2 atm.
4. Calculate the flue gas composition for the theoretical combustion of
   a. methane using 7% excess air,
   b. the dry gas composition case,
   c. the air–fuel ratio, and
   d. the flue gas quantity.
5. A fuel oil of specific gravity 0.90 contains 89% by weight carbon, 10% by weight hydrogen, 1% by weight sulfur. Calculate the dry and wet flue gas compositions for combustion in a theoretical amount of air.
6. Calculate the flue gas composition on a wet and dry basis and the quantity in A m^3/h at 170°C from a 200-MW boiler. The coal analysis is 82% C, 6% H_2, 1% S, 1% N_2, 6% ash, and 4% O_2. Assume that the boiler operates at 32% excess air and the heat value of the coal is 6.8 kcal/g coal. Assume theoretical combustion.

References

Bienstock, D., Brunn, L. W., Murphy, E. M., and Benson, H. E. (1958). Sulfur dioxide, its chemistry and removal from flue gases. U.S. Bur. of Mines Informat. Circ. 7836.

Castellan, G. W. (1983). "Physical Chemistry," 3rd Ed. Addison-Wesley, Reading, Massachusetts.

Cheremisinoff, P. N., and Young, R. A., eds. (1975). "Pollution Engineering Practice Handbook," Chap. 7. Ann Arbor Science, Ann Arbor, Michigan.

Eckert, J. S. (1970). No mystery in packed-bed design. *Oil Gas J.* **68**(34), 58–64.

EPA (1973–1978). "Compilation of Air Pollutant Emission Factors." U.S. EPA, AP-42 plus supplements 1–8.

Gleason, R. J. (October 1971). "Venturi-Type Contactor for Removal of SO_2 by the Limestone Wet-Scrubbing System Process." Cottrell Environmental Systems, EPA Project No. CES-116.

Goldmann, F., and Polanyi, M. (1928). Adsorption of vapors on charcoal and the thermal expansion of the liquified surface film. *Z. Phys. Chem. Stoechiom. Verwandtschaftsl.* **132**, 321–370.

Gottschlich, C. F. (1977). Combustion. *In* "Air Pollution" (A. C. Stern, ed.), 3rd Ed., Vol. IV. Academic Press, New York.

Hesketh, H. E. (1979). "Air Pollution Control." Ann Arbor Science, Ann Arbor, Michigan.

Johnstone, H. F., and Silox, H. E. (1947). Gas adsorption and humidification in cyclone spray towers. *Ind. Eng. Chem.* **39**, 808–817.

Kaminsky, W. (1983). Processes for flue gas desulfurization. *Chem. Ing. Tech.* **55**, 667–678 (in German).

Katz, M., and Cole, R. J. (1950). Recovery of sulphur compounds from atmospheric contaminants. *Ind. Eng. Chem.* **42**, 2258–2269.

Koehler, G. R., and Dobe, E. J. (1974). New England SO_2 control project final results. *EPA FGD Symp., Atlanta, Nov.*

Kohl, A. L., and Riesenfeld, F. C. (1960). "Gas Purification." McGraw-Hill, New York.

Leonards, G. (1984). Odor sampling and analysis. *In* "Handbook of Air Pollution Technology" (S. Calvert and H. M. Englund, eds.), Chap. 33. Wiley, New York.

Lunde, K. E. (1958). Performance of equipment for control of fluoride emissions. *Ind. Eng. Chem.* **50**, 293–298.

Nicklin, T., and Holland, B. H. (1963). "Cleaning of Coke Oven Gas by the Stretford Process," pp. 1–8. Coke Oven Mrs. Assoc., Cardiff.

Rees, R. L. L. (1953). Removal of oxides of sulphur from flue gases. *J. Inst. Fuel* **25**, 350–357.

Rolke, R. W., Hawthorne, R. D., Garbett, C. R., Slater, E. R., Phillips, T. T., and Towell, G. D. (1972). "Afterburner Systems Study." Report EPA-R2-72-062. U.S. Environmental Protection Agency, Washington, D.C.

Satriana, M., ed. (1981). "New Developments in Flue Gas Desulfurization Technology." Noyes Data Corp., Park Ridge, New Jersey.

Schrod, M., Semel, J., and Steine, R. (1981). *Chem. Ing. Tech.* **57**, 717 (in German).

Slack, A. V. (1975). Removal of sulfur oxides from stack gas: Recovery versus throway methods. *In* "Industrial Air Pollution Control" (K. Noll and J. Duncan, eds.), Chap. 8. Ann Arbor Science, Ann Arbor, Michigan.

Smith, W. S., and Gruber, C. W. (1966). "Atmospheric Emission from Coal Combustion—An Inventory Guide." U.S. Dept. of Health, Education and Welfare, Publ. #998-AP-24.

Strauss, W. (1975). "Industrial Gas Cleaning," 2nd Ed. Pergamon, Oxford.

Treybal, R. E. (1980). "Mass-Transfer Operations," 3rd Ed. McGraw-Hill, New York.

Turk, A. (1977). Adsorption. *In* "Air Pollution" (A. C. Stern, ed.), Vol. IV., Chap. 8. Academic Press, New York.

Turk, A., Sleik, H., and Messer, P. J. (1952). Determination of gaseous air pollution by carbon adsorption. *Am. Ind. Hyg. Assoc.* **13**, 23–28.

U.S. Stoneware (1967). "Design Information for Packed Towers." Norton Bull. DC-10.

Wheeler, A. (1951). "Advances in Catalysis" (W. G. Frankenburg *et al.*, eds.), Vol. 3, p. 250. Academic Press, New York.

Whynes, A. L. (1956). Some aspects of absorption of silicon tetrafluoride gas in water. *Trans. Inst. Chem. Eng.* **34**, 117–126.

9

Source Testing

9.1 Introduction

Source sampling for particulates (Sparks, 1984) and for gases (Knapp, 1984) in industrialized nations is regulated by law. It forms the basis for the determination of emission factors and the effectiveness of air pollution control equipment. Before it is possible to specify suitable air pollution control equipment it is necessary to know the following:

1. The gas flow rates
2. The gas composition
3. The gas temperature
4. The moisture content of the gases
5. The dew point of the gases
6. The particulate loading
7. The particle size distribution
8. The physical and chemical properties of the particles

9.2 Measurement of Gas Flow Rates

Gas flow rates are most commonly determined with the help of pitot tubes, because they are very handy. They do not require a long calming section in the dust as they measure local velocities. They are small, so they can be inserted through a minute hole in the side of the duct without necessitating plant shutdown. They do not cause any appreciable pressure loss in the gas stream.

In order to find the total gas flow rate, one must perform a pitot tube traverse. Circular ducts are divided into a number of annuli of equal area, and the velocity is measured on two diameters intersecting at right angles giving four velocities for each annulus. In rectangular ducts, the duct cross section is divided into a number of small rectangles, and the velocity is measured in the center of each.

The gas velocity at a single point, v_s, is calculated by the formula

$$v_s = K_p C_p \sqrt{\Delta p}\sqrt{T_s/M_{ws} p_s} \tag{9.1}$$

where $K_p = 85.58$, C_p is the pitot tube calibration factor, Δp is the measured (impact) velocity pressure (in H_2O), T_s is the temperature of the stack (or duct) gas (°R), M_{ws} is the molecular weight (wet basis) of the stack gas (lb/lb mol), p_s is the absolute pressure of the stack (or duct) (in. Hg), and v_s is the gas velocity (ft/s).

$$p_s = p_{\text{barometric}} + \frac{\text{duct static pressure}}{13.6} \tag{9.2}$$

where the dust static pressure is measured in inches of water.

$$M_{ws} = M_{wd}(1 - B_{wo}) + 18B_{wo} \tag{9.3}$$

where $M_{wd} = 28.9$ is the carrier gas molecular weight for dry air and B_{wo} is the proportion by volume of water vapor in the gas, expressed as a decimal fraction, that is, 10% moisture = 0.10.

$$B_{wo} = p_{H_2O} RH/p_s \tag{9.4}$$

where p_{H_2O} is the vapor pressure of water at the temperature of the gas being sampled (in. Hg) and RH is the relative humidity expressed as a decimal fraction, that is, 20% $RH = 0.20$.

The volumetric flow rate is calculated as follows. The average gas velocity, $\bar{v}_s$ is calculated by averaging the square roots of the measured Δp's and using Eq. (9.1). Then the gas volumetric flow rate (actual) in actual cubic feet per minute (ACFM) is obtained:

$$\text{ACFM} = \bar{v}_s 60 A_s \tag{9.5}$$

where A_s = cross-sectional area of duct or stack in square feet.

From this the gas volumetric flow rate at standard conditions (70°F and 29.92 in. Hg) on a dry basis is calculated:

$$\begin{array}{l}\text{DSCFM (dry standard}\\ \text{cubic feet per minute)}\end{array} = \text{ACFM}\frac{p_s}{p_{ref}}\frac{T_{ref}}{T_s}(1 - B_{wo}) \tag{9.6}$$

where p_{ref} is the reference (standard) pressure (29.92 in. Hg), T_{ref} is the reference (standard) temperature (530°R), and T_s is the absolute stack temperature (°R).

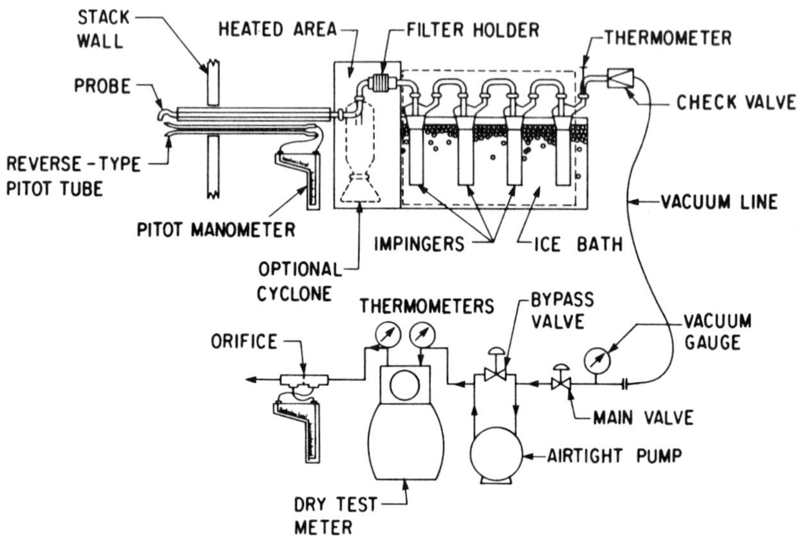

Figure 9.1 The EPA particulate sampling train (Sparks, 1984).

9.3 Measurement of Gas Composition

It is of paramount importance to ensure that the sample of the gas analyzed be representative of the total gas stream sampled. As shown in Fig. 9.1, a sample probe is inserted in the duct with the probe opening facing upstream. The gases entering the probe are passed through a sampling train. Essentially, the same type of sampling train is used for sampling gases and for sampling particulates. [For more information see, e.g., Knapp (1984) or Strauss (1975).]

One of the oldest and most widely used methods of gas analysis is absorption of the different constituent gases in solutions in the Orsat apparatus. Carbon dioxide is absorbed in caustic potash, oxygen in alkaline pyrogallol, and carbon monoxide in ammoniacal curpous chloride. The process has been mechanized in units such as the Mono Duplex (made by James Gordon and Co. Ltd.) where the initial carbon dioxide absorption and volume reduction is followed by combustion of unburned gases, that is, carbon monoxide and hydrogen, and a second carbon dioxide absorption. Density instruments are used, for example, to determine the carbon dioxide (CO_2) content in the air because of its higher density than the other main components of air. The density meter is usually a comparison-type instrument where two fast-moving impellers immersed in air and gas, respectively, are driven in opposite directions. Thermal conductivity is used in instruments built on the Wheatstone bridge principle. Carbon dioxide has about 60% of the thermal conductivity of air, while sulfur dioxide has only one-third of the thermal conductivity of air, so each of these two contaminants interferes with the analysis of the other. Infrared absorption can be used for routine simultaneous analysis of SO_2, CO, CO_2, H_2O, N_2O, CH_4, and organic vapors. Ultraviolet spectroscopy has also been used successfully for continuous monitoring of gases present in flue gases, including SO_2, NO_2, Cl_2, Br_2, F_2, H_2S, acetone, NH_3, etc. Oxygen is paramagnetic, while most common gases are diamagnetic. This property is utilized in an instrument based on the magnetic wind principle.

9.4 Measurement of Gas Temperature

When measuring the gas temperature, attention should be paid to errors of measurement whenever the temperature of the gas is different from that of the walls of the duct. Ordinary mercury-in-glass thermometers will cover the temperature range from below 0°C to about 350°C. An alcohol-in-glass thermometer extends the lower end of this range to − 80°C. Platinum resistance thermometers have a range from −200 to about +600°C. There are a variety of different thermocouples that collectively cover a range of temperatures from −200 to +2500°C.

9.5 Determination of the Moisture Content of the Gas

If a gas contains a vapor in such proportions that its partial pressure is less than the vapor pressure of the liquid at the existing temperature, the mixture is but partially saturated. The *relative saturation* of such a mixture is defined as the ratio of the partial pressure of the vapor to the vapor pressure of the liquid at the existing temperature. When the vapor under consideration is water, the relative saturation is termed *relative humidity*. Relative humidity can be used to calculate the moisture content of the gas by Eq. (9.4).

Relative humidity is usually determined by means of wet- and dry-bulb thermometry with the aid of a molal humidity chart. (Molal humidity is the number of moles of water per mole of moisture-free gas.) The principles of wet- and dry-bulb thermometry are as follows.

When a liquid evaporates into a large volume of an unsaturated gas, which is initially at the same temperature as the liquid, the liquid will spontaneously cool to supply part of the energy required for vaporization. If the body of the liquid has a large surface area in proportion to its mass, its temperature will quickly drop to an equilibrium value. This equilibrium temperature is called the wet-bulb temperature and is always less than the actual dry-bulb temperature of the gas into which evaporation is taking place. Provided that the heat of vaporizaton comes only from cooling of the ambient air by the liquid, the depression of the wet-bulb temperature is a measure of the degree of unsaturation of the mixture of gas and vapor, that is, of the relative humidity. The humidity charts establish (among other things) the relationship between dry-bulb temperature, wet-bulb temperature, and the relative humidity at a constant total pressure of 1.0 atm.

9.6 Determination of Dew Point of the Gas

If an unsaturated mixture of vapor and gas is cooled, the relative amounts of the components and the percentage composition by volume will at first remain unchanged. It follows that, if the total pressure is constant, the partial pressure of the vapor will not change on cooling. This will be the case until the temperature is lowered to such a value that the vapor pressure of the pure liquid at this temperature is equal to the existing particle pressure of the vapor in the mixture. The mixture will then be saturated, and any further cooling will result in condensation. The temperature at which the equilibrium vapor pressure of the liquid is equal to the existing partial pressure of the vapor is the *dew point* of the mixture. Provided that the equilibrium vapor pressure at the temperature of the gas is known, the moisture content of the gas may be calculated from the dew point. This naturally involves cooling a portion of the gas–vapor mixture. The accurate knowledge of the dew point of the gases to be cleaned is of great importance, for collection below the dew point takes place under wet conditions.

9.7 Determination of Particulate Loadings

The method of sampling for particulates from a gas stream is essentially the same as that of sampling for gases, but extra precautions must be taken to ensure a representative sample. This is done by performing "isokinetic" sampling, that is, keeping the sampling velocity the same as the duct velocity, as shown in Fig. 9.2(a). If the sample velocity is too low, the streamline pattern about the probe will appear as shown in Fig. 9.2(b). In this eventuality, some of the particles may leave the streamlines and enter the probe, which results in oversampling. However, if the sampling velocity is too high, as shown in Fig. 9.2(c), some of the particles may leave the streamlines and miss the probe, which results in undersampling. The

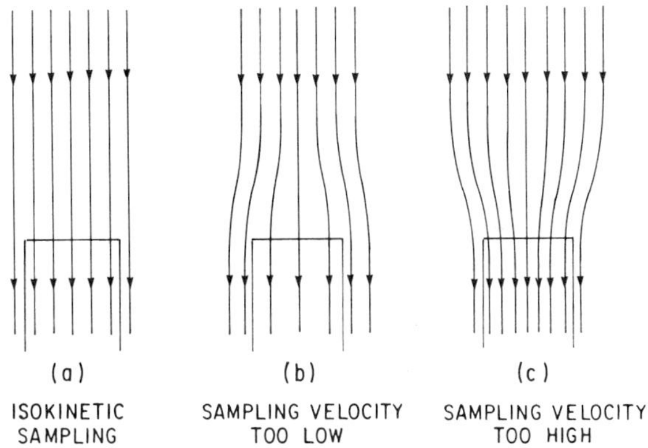

Figure 9.2 Gas streamlines at the entrance to sampling probes. [Reprinted with permission from Strauss (1975).]

percentage errors in the measured particulate loadings of 5- and 10-μm particles with a density of 2 g cm^{-3}, sampled at 7.6 m s^{-1}, are shown in Fig. 9.3 in dependence of the gas velocity. The error increases with the inertial impaction parameter Ψ.

The isokinetic sampling rate is calculated by the formula

$$\text{Des CFM} = 27.96 C_p \sqrt{\frac{T_s}{p_s M_{ws}}} (1 - B_{wo}) \frac{T_m}{T_s} \frac{p_s}{p_m} (Nd^2) \sqrt{\Delta p} \tag{9.7}$$

where Des CFM is the desired cubic feet per minute through the dry gas meter; C_p is the pitot tube calibration factor; T_s is the temperature of the stack (or duct) gas (°R); p_s is the absolute pressure of the stack (or duct) (in. Hg); M_{ws} is the molecular weight of the stack (or duct) gas on a wet basis (lb/lb mol); B_{wo} is the proportion by volume of water vapor in the gas stream expressed as a decimal fraction; T_m is the average temperature of the dry gas meter (°R); p_m is the absolute pressure of the dry gas meter (in. Hg); Nd is the average inside diameter of the sampling nozzle (in.); and Δp is the measured velocity pressure head (in. H_2O).

Membrane filters (e.g., Millipore Filter Co.) are the most widely used filters. They have holes which are very closely sized The most common grade used for aerosol filtration has holes 0.8 μm $\pm$ 0.05 μm, but filters with smaller holes are also in use. For the collection of larger particles, small cyclone collectors are sometimes used in the sampling train. When size classification of the particles is required as they are collected by the probe, a cascade impactor is connected to the probe (see Section 9.8).

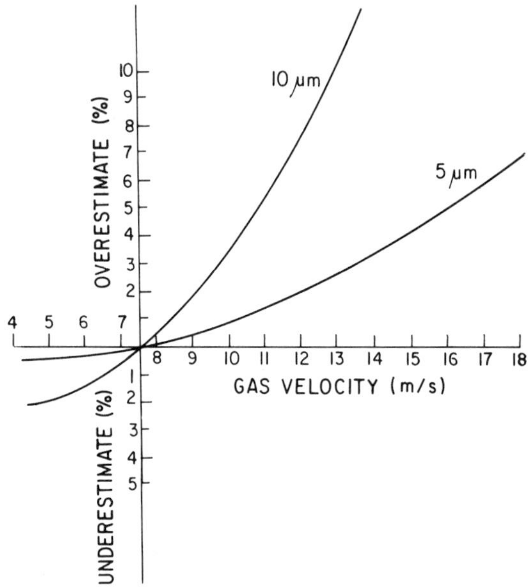

Figure 9.3 Errors in concentrations of particles of 5 and 10 μm sampled at 7.6 m s^{-1} with a 25-mm-diam nozzle at ambient conditions. [Reprinted with permission from Strauss (1975).]

The volume of the dry gas sampled is measured directly by the dry test meter. From this the volume of gas sampled at standard (reference) conditions, $V_{\mathrm{m\,ref}}$ (i.e., 70°F and 29.92 in. Hg, on a dry basis), is calculated as

$$V_{\mathrm{m\,ref}} = 17.71\left(\frac{V_{\mathrm{m}}}{T_{\mathrm{m}}}\right)\left(p_{\mathrm{barom}} + \frac{\Delta H}{13.6}\right) \tag{9.8}$$

where V_{m} is the volume of gas throughout the dry gas meter at meter conditions (ft^3); T_{m} is the average temperature of the gas meter (°R); p_{barom} is the average barometric pressure (in. Hg); H is the average pressure differential acros the orifice at the outlet of the dry gas meter (in. H_2O); and 17.71 = 530/29.92.

The dust loading, L, in grains per standard cubic foot is calculated as

$$L_{\mathrm{(grains/DSCF)}} = M_{\mathrm{p}} \times 15.43/V_{\mathrm{m\,ref}} \tag{9.9}$$

where M_{p} is the total particulate collected in the sampling train (grams).

9.8 Particle Size Distribution

There are a number of methods (e.g., microscopy, Coulter counter, etc.) to size dust particles after they have been collected on a filter paper or in a

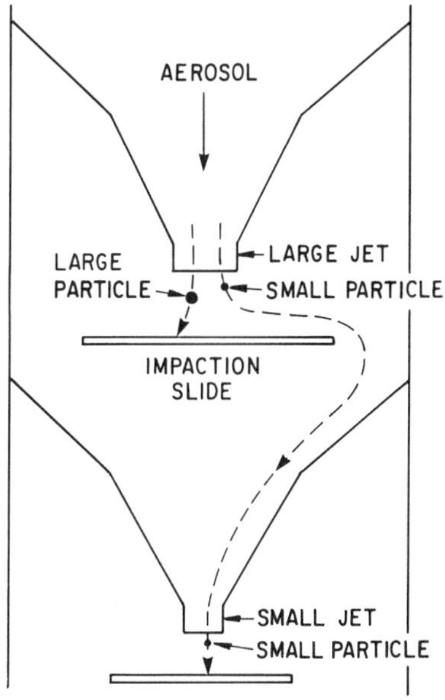

Figure 9.4 Principle of the cascade impactor. [Reprinted with permission from Strauss (1975).]

cyclone separator. All of these methods, however, have at least two serious limitations. First, most particles in the bulk state, that is, after collection, tend to form agglomerates that can be broken up into single particles only with considerable difficulty. As it is not known whether the particles, while carried by the gas stream, were single particles or agglomerates, these methods cannot be relied on to result in the particle sizes present in the gas stream. Second, the particle shape and the effective particle density are very difficult to determine, yet these quantities are necessary to obtain a representative value of the inertial impaction parameter.

Both of these limitations of particle sizing may be eliminated if aerodynamic particle sizing is used. Aerodynamic particle sizing is done with the help of cascade impactors. As shown diagrammatically in Fig. 9.4, a cascade impactor is constructed of a succession of jets, each followed by an impaction plate. It is based on the principle of inertial impaction. As each jet is smaller than the preceding one, the velocity of the air stream, and therefore that of the particles carried by the air stream, is increased as the aerosol advances through the impactor. Consequently, on the first impaction plate only the largest particles will impact, on the second impaction plate somewhat smaller particles will impact, and so forth. Before the gases exit from the cascade impactors, they pass through a backup filter capable of collecting particles as small as 0.05 μm. There are a number of commercially available cascade impactors, for example, the Andersen in-stack dust sampler. It is equipped with a probe, and one proceeds with the sampling in a manner similar to the case of ordinary particulate sampling. Usually, there are 9 jet plates, each having a pattern of precision drilled orifices. The 9 plates are separated by 2.5-mm stainless steel spacers. The jets on each plate are arranged in modified concentric circles that are offset on each succeeding plate. The size of the orifices is the same on a given plate, but is smaller for each succeeding downstream plate. Since the particle deposit areas are directly below the jets, 7 of the plates act as both a jet stage and a collection stage. The "0" stage is a jet stage only and the "8" stage is a collection stage only.

The cascade impactor is equipped with calibration charts that give the aerodynamic cut-off diameter—the diameter at which particle collection efficiency for the stage is 50%—for each stage as a function of the flow rate through the impactor. The aerodynamic diameter is defined as the diameter of a unit density sphere with the same settling velocity as the particle in question, that is,

$$d_{\mathrm{Ae}} = \sqrt{18\mu u_{\mathrm{t}}/gC(d_{\mathrm{Ae}})} \tag{9.10}$$

which is to be compared with the "Stokes diameter" d_{st}, defined by the equation

$$d_{\mathrm{st}} = \sqrt{18\mu u_{\mathrm{t}}/\rho gC(d_{\mathrm{st}})} \tag{9.11}$$

whence

$$d_{\mathrm{st}} = d_{\mathrm{Ae}}\sqrt{C(d_{\mathrm{Ae}})/\rho C(d_{\mathrm{st}})} \tag{9.12}$$

The last correction is also given as a function of ρ, in the form of a chart. It is valid only for spherical particles, and it assumes that ρ is accurately known. It is unfortunate that the calibration is not given in terms of the aerodynamic impaction diameter

$$d_{\mathrm{a}} = d_{\mathrm{st}}\sqrt{C(d_{\mathrm{st}})\rho} \tag{9.13}$$

which makes it unnecessary to know the particle density and shape, and yields information that can be used directly in calculating the inertial impaction parameter.

References

Knapp, K. T. (1984). Sampling and analysis of source gases. *In* "Handbook of Air Pollution Technology" (S. Calvert and H. M. Eglund, eds.), Chap. 29. Wiley, New York.

Sparks, L. E. (1984). Particulate sampling and analysis, Q.31. *In* "Handbook of Air Pollution Technology" (S. Calvert and H. M. Eglund, eds.). Wiley, New York.

Strauss, W. (1975). "Industrial Gas Cleaning." Pergamon, Oxford.

Appendix A

Conversion Factors and Physical Constants

Length

1 Ångström unit (Å) = 10^{-8} cm
1 ft = 30.48 cm
1 in. = 2.54 cm
1 m = 100 cm
1 μm = 10^{-6} m
1 mile (mi) = 5280 ft = 1609 m

Area

1 acre (A) = 4046.9 m^2
1 hectare (ha) = 2.471 acres
1 ft^2 = 929.0 cm^2
1 $in.^2$ = 6.452 cm^2
1 m^2 = 10.76 ft^2
1 mi^2 = 640 acres

Volume

1 ft^3 = 7.481 gallons (United States) = 0.02832 m^3
1 $in.^3$ = 16.39 cm^3
1 liter = 0.2642 gallon (United States)

Mass

1 grain = 0.0648 g
1 gram (g) = 10^{-3} kg
1 kg = 2.2046 lb
1 ton = 2000 lb
1 tonne = 1000 kg

Density

1 grain/ft^3 = 2.288 g/m^3
1 lb/ft^3 = 16.02 kg/m^3
1 g/cm^3 = 10^3 kg/m^3

Force

1 N = 10^5 dyn = 0.22481 lb$_f$

Pressure

1 atm = 1.01325 × 10^5 Pa = 101.325 kPa = 760 mm Hg = 1034 cm H_2O
= 1.01325 bar = 14.696 lb$_f$/in.^{2} (psia)
1 Pa ≡ 1 N/m^2
1 cm H_2O = 980.638 dyn/cm^2 = 98.0638 Pa
1 in. H_2O = 249.082 Pa

Energy

1 J = 10^7 erg = 9.86923 cm^3 atm = 0.239006 cal
= 5.12197 × 10^{-3} ft^3 psia = 0.737562 ft lb$_f$
= 9.47831 × 10^{-4} Btu = 2.778 × 10^{-7} kW h
= 3.725 × 10^{-7} hp h

Power

kW = 1.3410 hp = 4.4254 × 10^4 ft lb$_f$/min = 56.896 Btu/min
= 14.333 × 10^3 cal/min

Temperature

K = °C + 273.16
°F = 1.8°C + 32
°R = °F + 459.49 = 1.8 K

Surface Tension

1 mN/m = 1 dyn/cm = 1 erg/cm^2

Viscosity

1 Pa s^{-1} = 10 P = 0.020886 lb$_f$ s/ft^2 = 1.4503 × 10^{-4} lb$_f$ s/in.^{2}

Diffusivity

1 m^2/s = 10^4 cm^2/s

Values of Universal Gas Constant

R = 8.314 J/mol K = 82.06 cm^3 atm/mol K
= 1.987 cal/mol K = 1.986 Btu/lb mol R
= 0.7302 ft^3 atm/lb mol R = 10.73 ft^3 psia/lb mol R
= 1545 ft lb_f/lb mol R

Avogadro's Number

$N = 6.02 \times 10^{23}$ molecules/mol

Appendix B

Moisture Content of Air

CONVERSION DATA – MOISTURE CONTENT (WET WEIGHT AND BONE DRY BASIS)

WET BASIS FORMULA: $\frac{\text{Wet wt.} - \text{Bone Dry wt.}}{\text{Wet wt.}} \times 100$ = Moisture Content in % Wet Basis (W.W.B.)

DRY BASIS FORMULA: $\frac{\text{Wet wt.} - \text{Bone Dry wt.}}{\text{Bone dry wt.}} \times 100$ = Moisture Content in % Dry Basis (B.D.B.)

CONVERSION FORMULA:

Wet Basis to dry basis $\frac{100 \times \%\ \text{W.W.B.}}{100 - \%\ \text{W.W.B.}} = \%\ \text{B.D.B.}$

Dry Basis to wet basis $\frac{100 \times \%\ \text{B.D.B.}}{100 + \%\ \text{B.D.B.}} = \%\ \text{W.W.B.}$

CONVERSION TABLE – MOISTURE CONTENT – WET BASIS TO DRY BASIS

% W.W.B	% B.D.B.	% W.W.B.	% B.D.B.	% W.W.B.	% B.D.B.	% W.W.B.	% B.D.B.	% W.W.B.	% B.D.B.	% W.W.B.	% B.D.B.
.1	.1001	13	14.94	31	44.93	49	96.08	67	203.03	85	566.66
.5	.5025	14	16.28	32	47.06	50	100.	68	212.50	86	614.29
.9	.9082	15	17.65	33	49.25	51	104.08	69	222.58	87	669.23
1.00	1.01	16	19.05	34	51.51	52	108.33	70	233.33	88	733.33
1.25	1.2648	17	20.48	35	53.85	53	112.77	71	244.83	89	809.09
1.50	1.5228	18	21.95	36	56.25	54	117.39	72	257.14	90	900.00
1.75	1.7722	19	23.46	37	58.73	55	122.22	73	270.37	91	1011.11
2	2.04	20	25.00	38	61.29	56	127.27	74	284.62	92	1150.00
3	3.10	21	26.58	39	63.93	57	132.56	75	300.00	93	1328.57
4	4.16	22	28.20	40	66.66	58	138.09	76	316.66	94	1566.66
5	5.26	23	29.87	41	69.49	59	143.90	77	334.78	95	1900.00
6	6.38	24	31.58	42	72.42	60	150.00	78	354.54	96	2400.00
7	7.52	25	33.33	43	75.44	61	156.41	79	376.19	97	3233.33
8	8.70	26	35.13	44	78.57	62	163.16	80	400.00	98	4900.00
9	9.89	27	36.99	45	81.82	63	170.27	81	426.31	99	9900.00
10	11.11	28	38.88	46	85.19	64	177.77	82	455.55	100	X
11	12.35	29	40.85	47	88.68	65	185.71	83	488.24		
12	13.63	30	42.86	48	92.3	66	194.12	84	526.00		

CONVERSION TABLE – MOISTURE CONTENT – DRY BASIS TO WET BASIS

% B.D.B.	% W.W.B.	% B.D.B.	% W.W.B.	% B.D.B.	% W.W.B.	% B.D.B.	% W.W.B.	% B.D.B.	% W.W.B.	% B.D.B.	% W.W.B.
.1	.0999	20	16.66	44	30.55	68	40.48	92	47.92	180	64.29
.5	.4975	21	17.35	45	31.03	69	40.83	93	48.19	190	65.52
1.00	.99	22	18.03	46	31.51	70	41.18	94	48.45	200	66.67
1.25	1.2345	23	18.70	47	31.97	71	41.52	95	48.72	225	69.23
1.50	1.4778	24	19.35	48	32.43	72	41.86	96	48.98	250	71.43
1.75	1.7199	25	20.00	49	32.89	73	42.20	97	49.24	275	73.33
2	1.96	26	20.63	50	33.33	74	42.53	98	49.49	300	75.00
3	2.91	27	21.26	51	33.78	75	42.86	99	49.75	350	77.78
4	3.85	28	21.88	52	34.21	76	43.18	100	50.00	400	80.00
5	4.76	29	22.48	53	34.64	77	43.50	102	50.50	450	81.82
6	5.66	30	23.08	54	35.06	78	43.82	104	50.98	500	83.33
7	6.54	31	23.66	55	35.48	79	44.13	106	51.46	600	85.71
8	7.41	32	24.24	56	35.89	80	44.44	108	51.92	700	87.50
9	8.26	33	24.81	57	36.31	81	44.75	110	52.38	800	88.89
10	9.09	34	25.37	58	36.71	82	45.05	115	53.49	900	90.00
11	9.91	35	25.93	59	37.11	83	45.36	120	54.55	1000	90.91
12	10.71	36	26.47	60	37.50	84	45.65	125	55.56	2000	95.24
13	11.50	37	27.01	61	37.89	85	45.95	130	56.52	4000	97.56
14	12.28	38	27.54	62	38.27	86	46.24	135	57.45	8000	98.77
15	13.04	39	28.06	63	38.65	87	46.52	140	58.33	9999	99.01
16	13.79	40	28.57	64	39.02	88	46.81	145	59.18		
17	14.53	41	29.08	65	39.39	89	47.09	150	60.00		
18	15.25	42	29.58	66	39.76	90	47.37	160	61.54		
19	15.97	43	30.07	67	40.12	91	47.64	170	62.96		

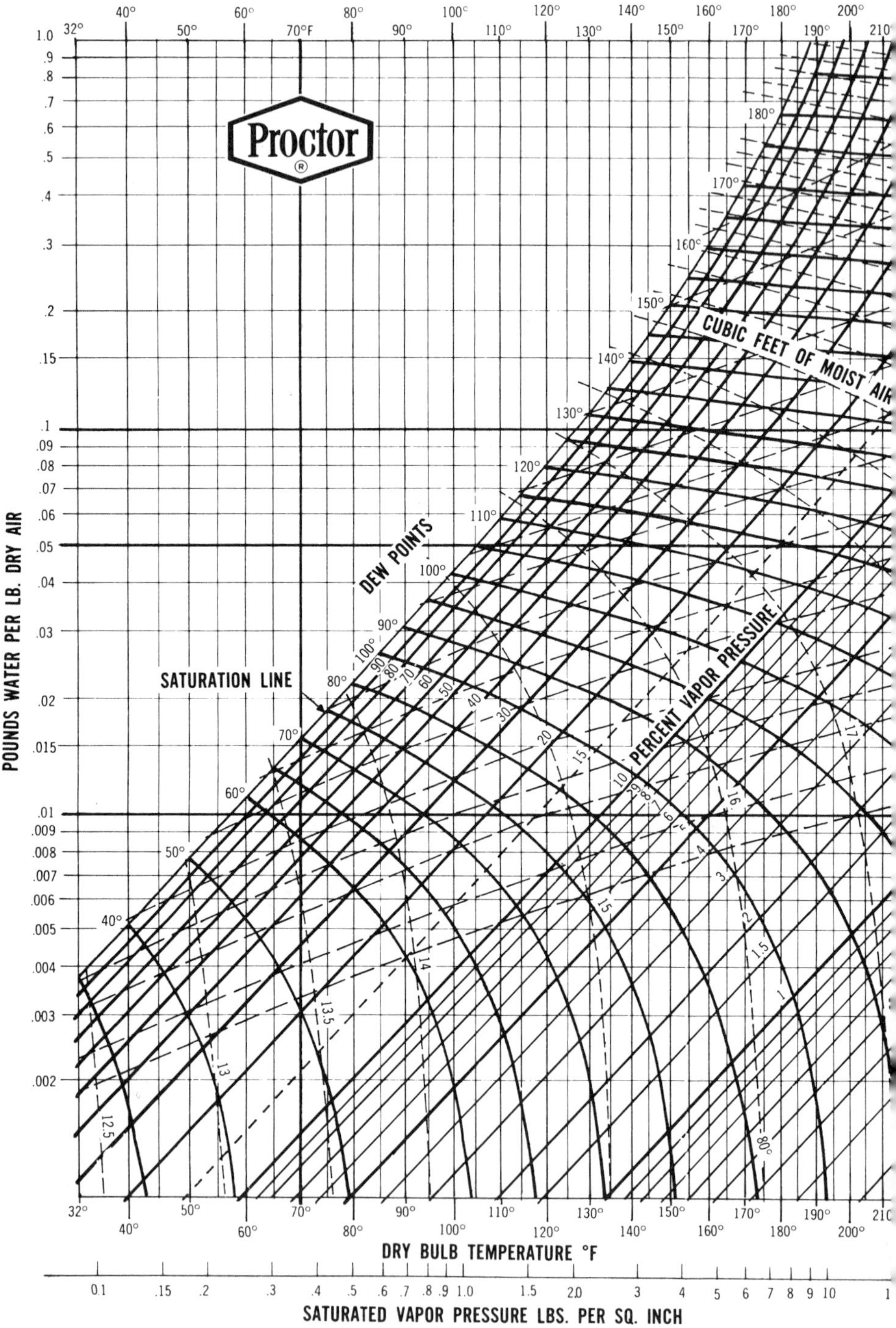
Proctor
POUNDS WATER PER LB. DRY AIR
DRY BULB TEMPERATURE °F
SATURATED VAPOR PRESSURE LBS. PER SQ. INCH
SATURATION LINE
DEW POINTS
PERCENT VAPOR PRESSURE
CUBIC FEET OF MOIST AIR

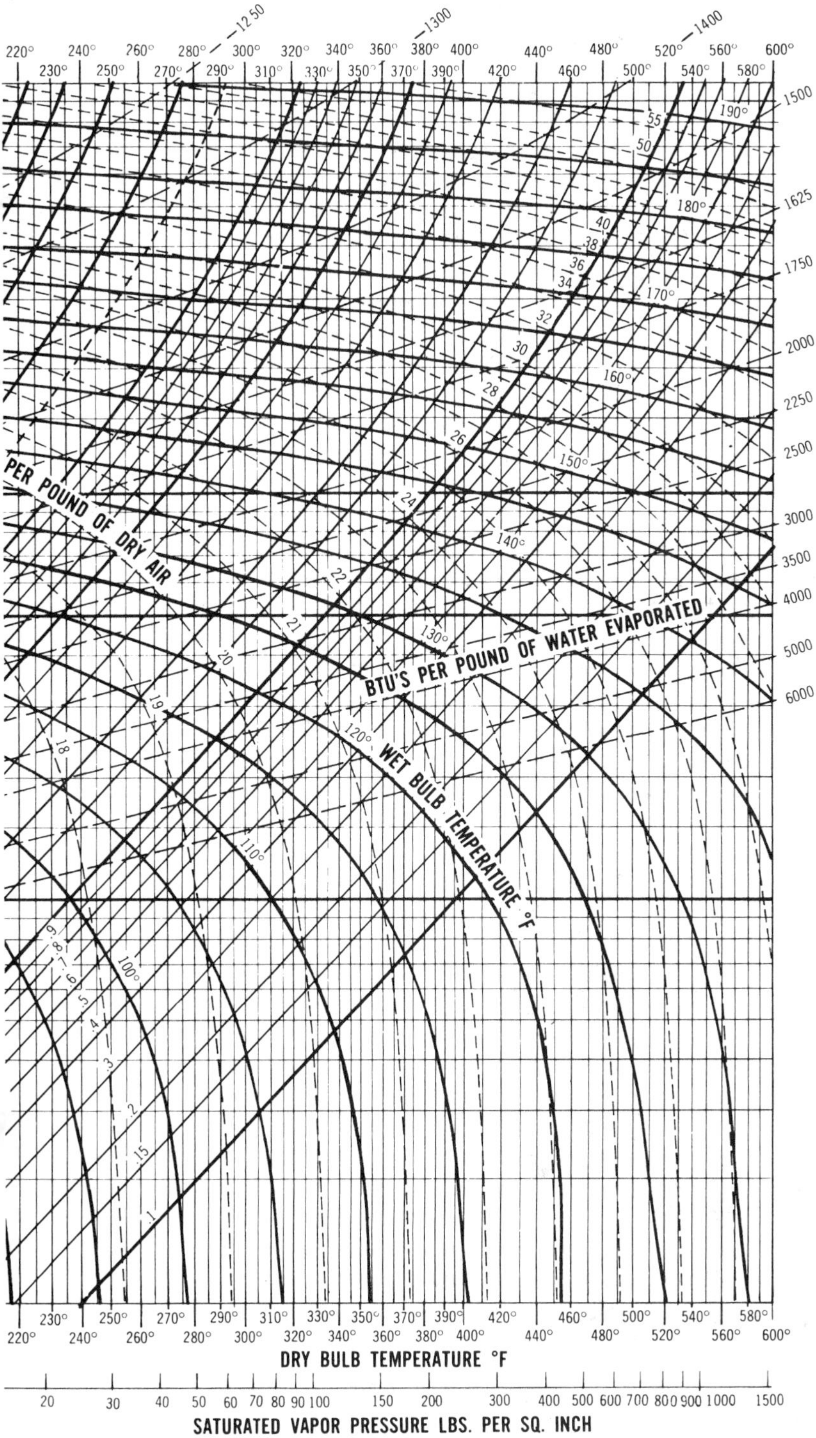

DEG.	DRY AIR CUBIC FEET PER POUND
32	12.392
35	12.469
40	12.594
45	12.723
50	12.837
55	12.970
60	13.089
65	13.228
70	13.333
75	13.477
80	13.605
85	13.717
90	13.850
95	13.986
100	14.104
105	14.225
110	14.368
115	14.493
120	14.620
125	14.749
130	14.859
135	14.993
140	15.106
145	15.221
150	15.337
155	15.504
160	15.625
165	15.748
170	15.873
175	16.000
180	16.129
185	16.260
190	16.367
195	16.502
200	16.639
205	16.750
210	16.892
215	17.007
220	17.123
225	17.271
230	17.391
235	17.513
240	17.637
245	17.794
250	17.921
255	18.018
260	18.149
265	18.282
270	18.416
275	18.519
280	18.657
285	18.797
290	18.939
295	19.048
300	19.194
305	19.305
310	19.417
315	19.569
320	19.685
325	19.802
330	19.920
335	20.040
340	20.161
345	20.325
350	20.408
355	20.576
360	20.704
365	20.790
370	20.921
375	21.053
380	21.186
385	21.322
390	21.413
395	21.552
400	21.692
405	21.834
410	21.930
415	22.075
420	22.173
425	22.321
430	22.422
435	22.573
440	22.676
445	22.831
450	22.936
455	23.095
460	23.202
465	23.310
470	23.474
475	23.529
480	23.697
485	23.810
490	23.981
495	24.096
500	24.213
505	24.331
510	24.450
515	24.570
520	24.691
525	24.876
530	24.938
535	25.063
540	25.189
545	25.316
550	25.445
555	25.575
560	25.707
565	25.804
570	25.974
575	26.110
580	26.178
585	26.385
590	26.455
595	26.596
600	26.738

Index